D1714318

Lecture Notes in Physics

Monographs

Springer
Berlin
Heidelberg
New York
Barcelona
Hong Kong
London
Milan
Paris
Singapore
Tokyo

http://www.springer.de/phys/

The Editorial Policy for Monographs

The series Lecture Notes in Physics reports new developments in physical research and teaching - quickly, informally, and at a high level. The type of material considered for publication in the monograph Series includes monographs presenting original research or new angles in a classical field. The timeliness of a manuscript is more important than its form, which may be preliminary or tentative. Manuscripts should be reasonably self-contained. They will often present not only results of the author(s) but also related work by other people and will provide sufficient motivation, examples, and applications.
The manuscripts or a detailed description thereof should be submitted either to one of the series editors or to the managing editor. The proposal is then carefully refereed. A final decision concerning publication can often only be made on the basis of the complete manuscript, but otherwise the editors will try to make a preliminary decision as definite as they can on the basis of the available information.
Manuscripts should be no less than 100 and preferably no more than 400 pages in length. Final manuscripts should be in English. They should include a table of contents and an informative introduction accessible also to readers not particularly familiar with the topic treated. Authors are free to use the material in other publications. However, if extensive use is made elsewhere, the publisher should be informed. Authors receive jointly 30 complimentary copies of their book. They are entitled to purchase further copies of their book at a reduced rate. No reprints of individual contributions can be supplied. No royalty is paid on Lecture Notes in Physics volumes. Commitment to publish is made by letter of interest rather than by signing a formal contract. Springer-Verlag secures the copyright for each volume.

The Production Process

The books are hardbound, and quality paper appropriate to the needs of the author(s) is used. Publication time is about ten weeks. More than twenty years of experience guarantee authors the best possible service. To reach the goal of rapid publication at a low price the technique of photographic reproduction from a camera-ready manuscript was chosen. This process shifts the main responsibility for the technical quality considerably from the publisher to the author. We therefore urge all authors to observe very carefully our guidelines for the preparation of camera-ready manuscripts, which we will supply on request. This applies especially to the quality of figures and halftones submitted for publication. Figures should be submitted as originals or glossy prints, as very often Xerox copies are not suitable for reproduction. For the same reason, any writing within figures should not be smaller than 2.5 mm. It might be useful to look at some of the volumes already published or, especially if some atypical text is planned, to write to the Physics Editorial Department of Springer-Verlag direct. This avoids mistakes and time-consuming correspondence during the production period.
As a special service, we offer free of charge LaTeX and TeX macro packages to format the text according to Springer-Verlag's quality requirements. We strongly recommend authors to make use of this offer, as the result will be a book of considerably improved technical quality.
For further information please contact Springer-Verlag, Physics Editorial Department II, Tiergartenstrasse 17, D-69121 Heidelberg, Germany.

Series homepage – http://www.springer.de/phys/books/lnpm

Ivan G. Avramidi

Heat Kernel and Quantum Gravity

Author

Ivan G. Avramidi
Department of Mathematics
New Mexico Institute of Mining and Technology
Socorro, NM 87801, USA

Library of Congress Cataloging-in-Publication Data applied for.

Die Deutsche Bibliothek - CIP-Einheitsaufnahme

Avramidi, Ivan G.:
Heat kernel and quantum gravity / Ivan G. Avramidi. - Berlin ; Heidelberg ; New York ; Barcelona ; Hong Kong ; London ; Milan ; Paris ; Singapore ; Tokyo : Springer, 2000
(Lecture notes in physics : N.s. M, Monographs ; 64)
ISBN 3-540-67155-2

ISSN 0940-7677 (Lecture Notes in Physics. Monographs)
ISBN 3-540-67155-2 Springer-Verlag Berlin Heidelberg New York

Springer-Verlag is a company in the BertelsmannSpringer publishing group.

Printed in Germany

Typesetting: Camera-ready by the author
Cover design: *design & production*, Heidelberg

Printed on acid-free paper
SPIN: 10644539 55/3144/du - 5 4 3 2 1 0

To my wife, Valentina, and my son, Grigori

Preface

This book is aimed primarily at theoretical physicists as well as graduate students working in quantum field theory, quantum gravity, gauge theories, and, to some extent, general relativity and cosmology. Although it is not aimed at a mathematically rigorous level, I hope that it may also be of interest to mathematical physicists and mathematicians working in spectral geometry, spectral asymptotics of differential operators, analysis on manifolds, differential geometry and mathematical methods in quantum theory. This book will certainly be considered too abstract by some physicists, but not detailed and complete enough by most mathematicians. This means, in particular, that the material is presented at the "physical" level of rigor. So, there are no lemmas, theorems and proofs and long technical calculations are omitted. Instead, I tried to give a detailed presentation of the basic ideas, methods and results. Also, I tried to make the exposition as explicit and complete as possible, the language less abstract and have illustrated the methods and results with some examples. As is well known, "one cannot cover everything", especially in an introductory text. The approach presented in this book goes along the lines (and is a further development) of the so-called background field method of De Witt. As a consequence, I have not dealt at all with manifolds with boundary, non-Laplace type (or nonminimal) operators, Riemann–Cartan manifolds as well as with many recent developments and more advanced topics, such as Ashtekar's approach, supergravity, strings, membranes, matrix models, M-theory etc. The interested reader is referred to the corresponding literature.

These lecture notes are based on my Ph.D. thesis at Moscow State University. Although most of the results presented here were published in a series of papers, this book allows for the much more detail and is easier to read. It can be used as a pedagogical introduction to quantum field theory and quantum gravity for graduate students with some basic knowledge of quantum field theory and general relativity. Based on this material, I gave a series of lectures for graduate students at the University of Naples during the fall semester of the 1995.

It should be noted that no attempts have been made to make the book completely self–consistent nor to give a fully comprehensive list of references. The bibliography reflects more or less adequately the situation of the late

1980s, when my original Ph.D. thesis was written. A complete update of the bibliography was obviously beyond my scope and capabilities. Nevertheless, I updated some old references and added some new ones that are intimately connected to the material of this book. I apologize in advance for not quoting the work of many authors who made significant contributions in the subject over the last decade. Besides, I believe that in an introductory text such as this a comprehensive bibliography is not as important as in a research monograph or a thorough survey.

I would like to express my sincere appreciation to many friends and colleagues who contributed in various ways to this book. First of all, I am especially indebted to Andrei O. Barvinsky, Vladislav R. Khalilov, Grigori M. Vereshkov and Grigori A. Vilkovisky who inspired my interest in quantum field theory and quantum gravity and from whom I learned most of the material of this book. I also have learned a great deal from the pioneering works of V.A. Fock, J. Schwinger and B.S. De Witt, as well as from more recent papers of T. Branson, S.M. Christensen, J.S. Dowker, M.J. Duff, E.S. Fradkin, S. Fulling, P.B. Gilkey, S. Hawking, H. Osborn, T. Osborn, L. Parker and A. Tseytlin among others. It was also a great pleasure to collaborate with Andrei Barvinsky, Thomas Branson, Giampiero Esposito and Rainer Schimming.

Over the last ten years my research has been financially supported in part by the Deutsche Akademische Austauschdienst, the Max Planck Institute for Physics and Astrophysics, the Russian Ministry for Science and Higher Education, Istituto Nazionale di Fizica Nucleare, the Alexander von Humboldt Foundation and the Deutsche Forschungsgemeinschaft.

Socorro, January 2000 *Ivan G. Avramidi*

Contents

Introduction 1

1. Background Field Method in Quantum Field Theory 9
1.1 Generating Functional, Green Functions and Effective Action 9
1.2 Green Functions of Minimal Differential Operators 14
1.3 Divergences, Regularization and Renormalization 17

2. Technique for Calculation of De Witt Coefficients 21
2.1 Covariant Expansions in Curved Space 21
2.2 Elements of Covariant Expansions 27
2.3 Technique for Calculation of De Witt Coefficients 34
2.4 De Witt Coefficients a_3 and a_4 37
2.5 Effective Action of Massive Fields 46

3. Partial Summation of Schwinger–De Witt Expansion 51
3.1 Summation of Asymptotic Expansions 51
3.2 Covariant Methods for Investigation of Nonlocalities 53
3.3 Summation of First-Order Terms 57
3.4 Summation of Second-Order Terms 61
3.5 De Witt Coefficients in De Sitter Space 68

4. Higher-Derivative Quantum Gravity 77
4.1 Quantization of Gauge Field Theories 77
4.2 One-Loop Divergences in Minimal Gauge 83
4.3 One-Loop Divergences in Arbitrary Gauge and Vilkovisky's Effective Action 94
4.4 Renormalization Group and Ultraviolet Asymptotics 101
4.5 Effective Potential 108

Conclusion 125

References 127

Notation 141

Index 143

Introduction

The classical macroscopic gravitational phenomena are described very well by the classical Einstein's general relativity [178, 211]. However, general relativity cannot be treated as a complete self-consistent theory in view of a number of serious difficulties that were not overcome since its creation [156].

This concerns, first of all, the problem of space-time singularities, which are unavoidable in the solutions of the Einstein equations [178, 211, 156, 151, 67]. In the vicinity of these singularities general relativity becomes incomplete as it cannot predict what is coming out from the singularity. In other words, the causal structure of the space-time breaks down at the singularities [151]. Another serious problem of general relativity is the problem of the energy of the gravitational field [100, 172, 70, 71].

The difficulties of the classical theory have motivated the need to construct a quantum theory of gravitation [99, 130]. Also the progress towards the unification of all non-gravitational interactions [128] shows the need to include gravitation in a general scheme of an unified quantum field theory.

The first problem in quantizing gravity is the construction of a covariant perturbation theory. Einstein's theory of gravitation is a typical non-Abelian gauge theory with the diffeomorphism group as a gauge group [78]. The quantization of gauge theories faces the known difficulty connected with the presence of constraints [207, 190]. This problem was successfully solved in the works of Feynman [104], De Witt [78] and Faddeev and Popov [101].

The most fruitful approach in quantum gravity is the background field method of De Witt [78], [80, 85]. This method is a generalization of the method of generating functionals in quantum field theory [50, 155, 193] to the case of non-vanishing background field. Both the gravitational field and the matter fields can have the background classical part.

The basic object in the background field method is the effective action functional. The effective action encodes, in principle, all the information of the standard quantum field theory. It determines the elements of the diagrammatic technique of perturbation theory, i.e., the full (or exact) one-point propagator and the full (or exact) vertex functions, with regard to all quantum corrections, and, hence, the perturbative S-matrix [78, 83, 223]. On the other hand, the effective action gives at once the physical amplitudes in real external classical fields and describes all quantum effects in external fields [81, 82] (vacuum polarization of quantized fields, particle creation etc.) [137, 42, 129, 187, 120, 121]. The effective action functional is the most appropriate tool for investigating the structure of the physical vacuum in various models of quantum field theory with spontaneous symmetry breaking (Higgs vacuum, gluon condensation, superconductivity) [191, 167, 164, 136, 53].

The effective action makes it possible to take into account the back-reaction of the quantum processes on the classical background, i.e., to obtain the effective equations for the background fields [83, 84, 223, 224, 122, 225, 113]. In this way, however, one runs into a difficulty connected with the dependence of the off-shell effective action on the gauge and the parametrization of the quantum field. In the paper [84] a gauge-invariant effective action (which still depends parametrically on the gauge fixing and the parametrization) was constructed. An explicitly reparametrization invariant functional that does not depend on the gauge fixing (so called Vilkovisky's effective action) was constructed in the papers [223, 224]. The "Vilkovisky's" effective action was studied in the paper [114] in different models of quantum field theory (including Einstein gravity) and in the paper of the author and Barvinsky [22] in case of higher-derivative quantum gravity. The Vilkovisky's effective action was improved further by De Witt in [86]. This effective action is called Vilkovisky–De Witt effective action. However, in many cases this modification has no effect on one–lop results, that is why we will not consider it in this book (for more details, see the original papers or the monograph [53]).

Thus, the calculation of the effective action is of high interest from the point of view of the general formalism as well as for concrete applications. The only practical method for the calculation of the effective action is the perturbative expansion in the number of loops [50, 155, 193]. All the fields are split in a background classical part and quantum perturbations propagating on this background. The part of the classical action, which is quadratic in quantum fields, determines the propagators of the quantum fields in background fields, and higher-order terms reproduce the vertices of the perturbation theory [83].

At one-loop level, the contribution of the gravitational loop is of the same order as the contributions of matter fields [93, 166]. At usual energies much lower than the Planck energy, $E_{\mathrm{Planck}} = \hbar c^5/G \approx 10^{19}\,\mathrm{GeV}$, the contributions of additional gravitational loops are highly suppressed. Therefore, a semi-classical concept applies when the quantum matter fields together with the linearized perturbations of the gravitational field interact with the background gravitational field (and, probably, with the background matter fields) [137, 42, 129, 187, 93, 166]. This approximation is known as one-loop quantum gravity [83, 150, 91, 72, 99, 130].

To evaluate the effective action it is necessary to find, first of all, the Green functions of the quantum fields in the background classical fields of different nature. The Green functions in background fields were investigated by a number of authors. Fock [105] proposed a method for solving the wave equation in background electromagnetic field by an integral transform in the proper time parameter (so called fifth parameter). Schwinger [203, 204] generalized the proper time method and applied it to the calculation of the one-loop effective action. De Witt [80, 82] reformulated the proper time method in geometrical language and applied it to the case of background gravitational field. Analogous questions for the elliptic partial differential operators were investigated by mathematicians (see the bibliography). In the papers [34, 35] the standard Schwinger–De Witt technique was generalized to the case of arbitrary differential operators satisfying the condition of causality.

The proper time method gives at once the Green functions in the neighborhood of the light-cone. Therefore, it is the most suitable tool for investigation of the ultraviolet divergences (calculation of counter-terms, β-functions and anomalies). The most essential advantage of the proper time method is that it is explicitly covariant and enables one to introduce various covariant regularizations of divergent integrals. The most popular are the analytical regularizations: dimensional regularization, ζ-function regularization etc. [137, 42, 97]. There are a lot of works along this line of investigation over the last two decades (see the bibliography). Although most of the papers restrict themselves to the one-loop approximation, the proper time method is applicable at higher loops too. In the papers [157, 169, 61, 36, 59] it was applied to analyze two-loop divergences in various models of quantum field theory including Einstein's quantum gravity.

Another important area, where the Schwinger–De Witt proper-time method is successfully applied, is the vacuum polarization of massive quantum fields by background fields. When the Compton wave length $\lambda = \hbar/mc$, corresponding to the field mass m, is much smaller than the characteristic length scale L of the background field, the proper time method gives immediately the expansion of the effective action in a series in the small parameter $(\lambda/L)^2$ [120, 121, 218]. The coefficients of this expansion are proportional to the so-called De Witt coefficients and are local invariants, constructed from the background fields and their covariant derivatives. In the papers [119, 223] the general structure of the Schwinger–De Witt expansion of the effective action was discussed. It was pointed out that there is a need to go beyond the limits of the local expansion by the summation of the leading derivatives of the background fields in this expansion. In the paper [223], based on some additional assumptions concerning the convergence of the corresponding series and integrals, the leading derivatives of the background fields were summed up and a non-local expression for the one-loop effective action in case of a massless field was obtained.

Thus, so far, effective and manifestly covariant methods for calculation of the effective action in arbitrary background fields are absent. All the calculations performed so far concern either the local structures of the effective action or some specific background fields (constant fields, homogeneous spaces etc.) [137, 42].

That is why the development of general methods for covariant calculations of the effective action, which is especially needed in quantum theory of gauge fields and gravity, is an actual and new area of research. There are many papers (see, among others, [8]–[40]), which are devoted to the development of this line of investigation. Therein an explicitly covariant and effective technique for the calculation of De Witt coefficients is elaborated. This technique is applicable in the most general case of arbitrary background fields and spaces and can be easily adopted to automated symbolic computation on computers [44]. In the papers [6, 12] the renormalized one-loop effective action for massive scalar, spinor and vector fields in background gravitational field up to terms of order $O(1/m^6)$ is calculated.

In spite of impressive progress in one-loop quantum gravity, a complete self-consistent quantum theory of gravitation does not exist at present [154]. The difficulties of quantum gravity are connected, in the first line, with the fact that there is no consistent way to eliminate the ultraviolet divergences arising in perturbation theory [229, 95]. It was found [60, 212, 163, 54, 131, 66, 94] that in the one-loop approximation the pure Einstein gravity is finite on mass shell (or renormalizable in case of non-vanishing cosmological constant). However, two-loop Einstein gravity is no longer renormalizable on-shell [135]. On the other hand, the interaction with the matter fields also leads to non-renormalizability on mass shell even in one-loop approximation [56, 58, 76, 75, 74, 184, 77, 206, 222, 33].

Among various approaches to the problem of ultraviolet divergences in quantum gravity (such as supergravity [221, 174, 103], resummation [79, 165, 197] etc. [229, 95]) an important place is occupied by the modification of the gravitational Lagrangian by adding quadratic terms in the curvature of general type (higher-derivative theory of gravitation). This theory was investigated by various authors both at the classical and at the quantum levels (see the bibliography).

The main argument against higher-derivative quantum gravity is the presence of ghosts in the linearized perturbation theory on flat background, that breaks down the unitarity of the theory [185, 186, 180, 208, 209, 160, 214, 213, 198, 145, 2, 170]. There were different attempts to solve this problem by the summation of radiative corrections in the propagator in the momentum representation [214, 213, 198, 145], [108, 109, 111]. However, at present they cannot be regarded as convincing in view of causality violation, which results from the unusual analytic properties of the S-matrix. It seems that the problem of unitarity can be solved only beyond the limits of perturbation theory [215].

Ultraviolet behavior of higher-derivative quantum gravity was studied in many papers (see the bibliography). However, the one-loop counter-terms were first obtained in the paper of Julve and Tonin [160]. The most detailed investigation of the ultraviolet behavior of higher-derivative quantum gravity was carried out in the papers of Fradkin and Tseytlin [108, 109, 111, 110]. In these papers, an inconsistency was found in the calculations of Julve and Tonin. The one-loop counter-terms were recalculated in higher-derivative quantum gravity of general type as well as in conformally invariant models and in conformal supergravity [111, 110]. The main conclusion of the papers [108, 109, 111, 110] is that higher-derivative quantum gravity is asymptotically free in the physical region of coupling constants, which is characterized by the absence of tachyons on the flat background. The presence of reasonably arbitrary matter does not affect this conclusion.

Thus, the investigation of the ultraviolet behavior of higher-derivative quantum gravity is an important and actual problem in the general program of constructing a consistent quantum gravity. It is this problem that was studied in the papers [22, 7]. Therein the off-shell one-loop divergences of higher-derivative quantum gravity in arbitrary covariant gauge of the quantum field were calculated. It was shown that the results of previous authors contain a numerical error in the coefficient of the R^2-divergent term. The correction of this mistake radically changed the asymptotic properties of the theory in the conformal sector. Although the conclusion of [108, 42, 111, 110] about the asymptotic freedom in the tensor sector of the theory remains true, the conformal sector exhibits just the opposite "zero-charge" behavior in the physical region of coupling constants considered in all previous papers (see the bibliography). In the unphysical region of coupling constants, which corresponds to the positive definiteness of the part of the Euclidean action quadratic in curvature, the zero-charge singularities at finite energies are absent.

The present book is devoted to further development of the covariant methods for calculation of the effective action in quantum field theory and quantum gravity, and to the investigation of the ultraviolet behavior of higher-derivative quantum gravity.

In Chap. 1. the background field method is presented. Sect. 1.1 contains a short functional formulation of quantum field theory in the form that is convenient for subsequent discussion. In Sect. 1.2 the standard proper time method with some extensions is presented in detail. Sect. 1.3 is concerned with the questions connected with the problem of ultraviolet divergences, regularization, renormalization and the renormalization group.

In Chap. 2 a manifestly covariant technique for the calculation of the De Witt coefficients is elaborated. In Sect. 2.1 the methods of covariant expansions of arbitrary fields in curved space with arbitrary linear connection in the generalized covariant Taylor series and the Fourier integral are formulated in the most general form. In Sect. 2.2 all the quantities that will be needed later are calculated in form of covariant Taylor series. In Sect. 2.3, based on the method of covariant expansions, the covariant technique for the calculation of the De Witt coefficients in matrix terms is developed. The corresponding diagrammatic formulation of this technique is given. The developed technique enables one to compute explicitly the De Witt coefficients as well as to analyze their general structure. The possibility to use the corresponding symbolic manipulations on computers is pointed out. In Sect. 2.4 the calculation of the De Witt coefficients a_3 and a_4 at coinciding points is presented. In Sect. 2.5. the one-loop effective action for massive scalar, spinor and vector fields in an background gravitational field is calculated up to terms of order $1/m^4$.

In Chap. 3 the general structure of the Schwinger–De Witt asymptotic expansion is analyzed and partial summation of various terms is carried out. In Sect. 3.1 a method for summation of the asymptotic series due to Borel (see, e.g., [192], sect. 11.4) is presented and its application to quantum field theory is discussed. In Sect. 3.2. the covariant methods for investigations of the non-local structure of the effective action are developed. In Sect. 3.3 the terms of first order in the background fields in De Witt coefficients are calculated and their summation is carried out. The non-local expression for the Green function at coinciding points, up to terms of second order in background fields, is obtained. The massless case is considered too. It is shown that in the conformally invariant case the Green function at coinciding points is finite at first order in background fields. In Sect. 3.4. the De Witt coefficients at second order in background fields are calculated. The summation of the terms quadratic in background fields is carried out, and the explicitly covariant non-local expression for the one-loop effective action up to terms of third order in background fields is obtained. All the formfactors, their ultraviolet asymptotics and imaginary parts in the pseudo-Euclidean region above the threshold are obtained explicitly. The massless case in four- and two-dimensional spaces is studied too. In Sect. 3.5 all terms without covariant derivatives of the background fields in De Witt coefficients, in the case of scalar field, are picked out. It is shown that in this case the asymptotic series of the covariantly constant terms diverges. By making use of the Borel summation procedure of the asymptotic series, the Borel sum of the corresponding semi-classical series is calculated. An explicit expression for the one-loop effective action, non-analytic in the background fields, is obtained up to the terms with covariant derivatives of the background fields.

Chapter 4 is devoted to the investigation of higher-derivative quantum gravity. In Sect. 4.1 the standard procedure of quantizing the gauge theories as well as the formulation of the Vilkovisky's effective action is presented. In Sect. 4.2 the one-loop divergences of higher-derivative quantum gravity with the help of the methods of the generalized Schwinger–De Witt technique are calculated. The error in the coefficient of the R^2-divergent term, due to previous authors, is pointed out. In Sect. 4.3 the dependence of the divergences of the effective action on the gauge of the quantum field is analyzed. The off-shell divergences of the standard effective action in arbitrary covariant gauge, and the divergences of the Vilkovisky's effective action, are calculated. In Sect. 4.4 the corresponding renormalization-group equations are solved and the ultraviolet asymptotics of the coupling constants are obtained. It is shown that in the conformal sector of the theory there is no asymptotic freedom in the "physical" region of the coupling constants. The presence of the low-spin matter fields does not change this general conclusion: higher-derivative quantum gravity necessarily goes beyond the limits of the weak conformal coupling at high energies. The physical interpretation of such ultraviolet behavior is discussed. It is shown that the asymptotic freedom both in tensor and conformal sectors is realized in the "unphysical" region of coupling constants, which corresponds to the positive-definite Euclidean action. In Sect. 4.5 the effective potential (i.e., the effective action on the De Sitter background) in higher-derivative quantum gravity is calculated. The determinants of the second- and fourth-order operators are calculated with the help of the technique of the generalized ζ-function. It is maintained that the result for the R^2-divergence obtained in Sect. 4.2, as well as the results for the arbitrary gauge and for the "Vilkovisky's" effective action obtained in Sect. 4.3, are correct. Both the effective potential in arbitrary gauge and the "Vilkovisky's" effective potential are calculated. The "Vilkovisky's" effective equations for the background field, i.e., for the curvature of De Sitter space, that do not depend on the gauge and the parametrization of the quantum field, are obtained. The first quantum correction to the background curvature caused by the quantum effects is found. In Conclusion the main results are summarized.

1. Background Field Method in Quantum Field Theory

1.1 Generating Functional, Green Functions and Effective Action

Let us consider an arbitrary field $\varphi(x)$ on a n-dimensional space-time given by its contravariant components $\varphi^A(x)$ that transform with respect to some (in general, reducible) representation of the diffeomorphism group, i.e. the group of general transformations of the coordinates. The field components $\varphi^A(x)$ can be of both bosonic and fermionic nature. The fermionic components are treated as anticommuting Grassmanian variables [41], i.e.,

$$\varphi^A\varphi^B = (-1)^{AB}\varphi^B\varphi^A , \qquad (1.1)$$

where the indices in the exponent of the (-1) are equal to 0 for bosonic indices and to 1 for the fermionic ones.

For the construction of a local action functional $S(\varphi)$ one also needs a metric of the configuration space E_{AB}, i.e., a scalar product

$$(\varphi_1, \varphi_2) = \varphi_1^A E_{AB} \varphi_2^B , \qquad (1.2)$$

that enables one to define the covariant fields components

$$\varphi_A = \varphi^B E_{BA} , \qquad \varphi^B = \varphi_A E^{-1\ AB} , \qquad (1.3)$$

where $E^{-1\ AB}$ is the inverse matrix

$$E^{-1\ AB} E_{BC} = \delta^A_{\ C} , \qquad E_{AC} E^{-1\ CB} = \delta_A^{\ B} . \qquad (1.4)$$

The metric E_{AB} must be non-degenerate both in bose-bose and fermi-fermi sectors and satisfy the supersymmetry conditions

$$E_{AB} = (-1)^{A+B+AB} E_{BA} , \qquad E^{-1\ AB} = (-1)^{AB} E^{-1\ BA} . \qquad (1.5)$$

In the case of gauge-invariant field theories we assume that the corresponding ghost fields are included in the set of the fields φ^A and the action $S(\varphi)$ is modified by inclusion of the gauge fixing and the ghost terms. To reduce the writing we will follow, hereafter, the condensed notation of De Witt [80, 83] and substitute the mixed set of indices (A, x), where x labels the space-time point, by one small Latin index $i \equiv (A, x)$: $\varphi^i \equiv \varphi^A(x)$. The combined *summation-integration* should be done over the repeated upper and lower small Latin indices

$$\varphi_{1\,i}\varphi_2^i \equiv \int \mathrm{d}^n x\, \varphi_{1\,A}(x)\varphi_2^A(x)\,. \tag{1.6}$$

Now let us single out two causally connected in- and out-regions in the space-time, that lie in the past and in the future respectively relative to the region, which is of interest from the dynamical standpoint. Let us define the vacuum states $|\mathrm{in}, \mathrm{vac} >$ and $|\mathrm{out}, \mathrm{vac} >$ in these regions and consider the vacuum-vacuum transition amplitude

$$\langle \mathrm{out}, \mathrm{vac} | \mathrm{in}, \mathrm{vac} \rangle \equiv \exp\left\{\frac{\mathrm{i}}{\hbar} W(J)\right\} \tag{1.7}$$

in presence of some background classical sources J_i vanishing in in- and out-regions.

The amplitude (1.7) can be expressed in form of a formal functional integral (or path integral) [50, 155, 193]

$$\exp\left\{\frac{\mathrm{i}}{\hbar} W(J)\right\} = \int \mathrm{d}\varphi\, \mathcal{M}(\varphi) \exp\left\{\frac{\mathrm{i}}{\hbar}\left[S(\varphi) + J_i\varphi^i\right]\right\}\,, \tag{1.8}$$

where $\mathcal{M}(\varphi)$ is a measure functional, which should be determined by the canonical quantization of the theory [117, 223]. The integration in (1.8) should be taken over all fields satisfying the boundary conditions determined by the vacuum states $|\mathrm{in}, \mathrm{vac} >$ and $|\mathrm{out}, \mathrm{vac} >$. The functional $W(J)$ is of central interest in quantum field theory. It is the generating functional for the Schwinger averages

$$\langle \varphi^{i_1} \cdots \varphi^{i_k} \rangle = \exp\left\{-\frac{\mathrm{i}}{\hbar} W(J)\right\} \left(\frac{\hbar}{\mathrm{i}}\right)^k \frac{\delta_L^k}{\delta J_{i_1} \cdots \delta J_{i_k}} \exp\left\{\frac{\mathrm{i}}{\hbar} W(J)\right\}\,, \tag{1.9}$$

where

$$\langle F(\varphi) \rangle \equiv \frac{\langle \mathrm{out}, \mathrm{vac} | \mathrm{T}\left(F(\varphi)\right) | \mathrm{in}, \mathrm{vac} \rangle}{\langle \mathrm{out}, \mathrm{vac} | \mathrm{in}, \mathrm{vac} \rangle}\,, \tag{1.10}$$

δ_L is the left functional derivative and 'T' is the operator of chronological ordering.

The first derivative of the functional $W(J)$ gives the mean field (according to the tradition we will call it the background field)

$$\langle \varphi^i \rangle \equiv \Phi^i(J) = \frac{\delta_L}{\delta J_i} W(J)\,, \tag{1.11}$$

the second derivative determines the one-point propagator

$$\langle \varphi^i \varphi^k \rangle = \Phi^i \Phi^k + \frac{\hbar}{\mathrm{i}} \mathcal{G}^{ik} \ , \tag{1.12}$$

$$\mathcal{G}^{ik}(J) = \frac{\delta_L^2}{\delta J_i \delta J_k} W(J) \ ,$$

and the higher derivatives give the many-point Green functions

$$\mathcal{G}^{i_1 \cdots i_k}(J) = \frac{\delta_L^k}{\delta J_{i_1} \cdots \delta J_{i_k}} W(J) \ . \tag{1.13}$$

The generating functional for the vertex functions, called the effective action $\Gamma(\Phi)$, is defined by the functional Legendre transform:

$$\Gamma(\Phi) = W(J) - J_i \Phi^i, \tag{1.14}$$

where the sources are expressed in terms of the background fields, $J = J(\Phi)$, by inversion of the functional equation $\Phi = \Phi(J)$, (1.11).

The first derivative of the effective action gives the sources

$$\frac{\delta_R}{\delta \Phi^i} \Gamma(\Phi) \equiv \Gamma_{,i}(\Phi) = -J_i(\Phi) \ , \tag{1.15}$$

the second derivative determines the one-point propagator

$$\frac{\delta_L \delta_R}{\delta \Phi^i \delta \Phi^k} \Gamma(\Phi) \equiv \mathcal{D}_{ik}(\Phi) \ , \qquad \mathcal{D}_{ik} = (-1)^{i+k+ik} \mathcal{D}_{ki} \ , \tag{1.16}$$

$$\mathcal{D}_{ik} \mathcal{G}^{kn} = -\delta_i{}^n \ ,$$

where δ_R is the right functional derivative, $\delta_i{}^n = \delta_A{}^B \delta(x, x')$, and $\delta(x, x')$ is the delta-function. The higher derivatives determine the vertex functions

$$\Gamma_{i_1 \cdots i_k}(\Phi) = \frac{\delta_R^k}{\delta \Phi^{i_1} \cdots \delta \Phi^{i_k}} \Gamma(\Phi) \ . \tag{1.17}$$

From the definition (1.14) and the equation (1.8) it is easy to obtain the functional equation for the effective action

$$\exp\left\{ \frac{\mathrm{i}}{\hbar} \Gamma(\Phi) \right\} = \int \mathrm{d}\varphi \, \mathcal{M}(\varphi) \exp\left\{ \frac{\mathrm{i}}{\hbar} \left[S(\varphi) - \Gamma_{,i}(\Phi)(\varphi^i - \Phi^i) \right] \right\} \ . \tag{1.18}$$

Differentiating the equation (1.15) with respect to the sources one can express all the many-point Green functions (1.13) in terms of the vertex functions (1.17) and the one-point propagator (1.12). If one uses the diagrammatic technique, where the propagator is represented by a line and the vertex functions by vertexes, then each differentiation with respect to the sources adds a new line in previous diagrams in all possible ways. Therefore, a many-point Green function is represented by all kinds of tree diagrams with a given number of external lines.

Thus when using the effective action functional for the construction of the S-matrix (when it exists) one needs only the tree diagrams, since all quantum corrections determined by the loops are already included in the full one-point propagator and the full vertex functions. Therefore, the effective equations (1.15),

$$\Gamma_{,i}(\Phi) = 0 \,, \tag{1.19}$$

(in absence of classical sources, $J = 0$) describe the dynamics of the background fields with regard to all quantum corrections.

The possibility to work directly with the effective action is an obvious advantage. First, the effective action contains all the information needed to construct the standard S-matrix [78, 161, 223]. Second, it gives the effective equations (1.19) that enable one to take into account the influence of the quantum effects on the classical configurations of the background fields [122, 225].

In practice, the following difficulty appears on this way. The background fields, as well as all other Green functions, are not Vilkovisky'sly defined objects. They depend on the parametrization of the quantum field [223, 224]. Accordingly, the effective action is not Vilkovisky's too. It depends essentially on the parametrization of the quantum field off mass shell, i.e., for background fields that do not satisfy the classical equations of motion

$$S_{,i}(\Phi) = 0 \,. \tag{1.20}$$

On mass shell, (1.20), the effective action is a well defined quantity and leads to the correct S-matrix [78, 161, 138, 155].

A possible way to solve this difficulty was proposed in the papers [223, 224], where an effective action functional was constructed, that is explicitly invariant with respect to local reparametrizations of quantum fields (so called Vilkovisky's effective action). This was done by introducing a metric and a connection in the configuration space. Therein, [223, 224], the "Vilkovisky's" effective action for the gauge field theories was constructed too. We will study the consequences of such a definition of the effective action in Chap. 5 when investigating the higher-derivative quantum gravity. This aproach was improved further by De Witt [86].

The formal scheme of quantum field theory, described above, begins to take on a concrete meaning in the framework of perturbation theory in the number of loops [50, 155, 193] (i.e., in the Planck constant $\hbar$):

$$\Gamma(\Phi) = S(\Phi) + \sum_{k\geq 1} \hbar^k \Gamma_{(k)}(\Phi) \,. \tag{1.21}$$

Substituting the expansion (1.21) in (1.18), shifting the integration variable in the functional integral $\varphi^i = \Phi^i + \sqrt{\hbar} h^i$, expanding the action $S(\varphi)$ and the measure $\mathcal{M}(\varphi)$ in quantum fields h^i and equating the coefficients at equal powers of $\hbar$, we get the recurrence relations that uniquely define all the coefficients $\Gamma_{(k)}$. All the functional integrals are Gaussian and can be calculated in the standard way [191]. As the result the diagrammatic technique for the effective action is reproduced. The elements of this technique are the bare one-point propagator, i.e., the Green function of the differential operator

$$\Delta_{ik}(\varphi) = \frac{\delta_L \delta_R}{\delta\varphi^i \delta\varphi^k} S(\varphi) \,, \tag{1.22}$$

and the local vertexes, determined by the classical action $S(\varphi)$ and the measure $\mathcal{M}(\varphi)$.

In particular, the one-loop effective action has the form

$$\Gamma_{(1)}(\Phi) = -\frac{1}{2\mathrm{i}} \log \frac{\operatorname{sdet} \Delta(\Phi)}{\mathcal{M}^2(\Phi)} \,, \tag{1.23}$$

where

$$\operatorname{sdet} \Delta = \exp\left(\operatorname{str} \log \Delta\right) \tag{1.24}$$

is the functional Berezin superdeterminant [41], and

$$\operatorname{str} F = (-1)^i F^i{}_i = \int \mathrm{d}^n x \, (-1)^A F^A{}_A(x) \tag{1.25}$$

is the functional supertrace.

The local functional measure $\mathcal{M}(\varphi)$ can be taken in the form of the superdeterminant of the metric of the configuration space

$$\mathcal{M} = \left(\operatorname{sdet} E_{ik}(\varphi)\right)^{1/2} \,, \tag{1.26}$$

where

$$E_{ik}(\varphi) = E_{AB}(\varphi(x))\delta(x, x') \,. \tag{1.27}$$

In this case $d\varphi\mathcal{M}(\varphi)$ is the volume element of the configuration space that is invariant under the point transformations of the fields: $\varphi(x) \to F(\varphi(x))$. Using the multiplicativity of the superdeterminant [41], the one-loop effective action with the measure (1.26) can be rewritten in the form

$$\Gamma_{(1)}(\Phi) = -\frac{1}{2\mathrm{i}} \log \operatorname{sdet} \hat{\Delta} \,, \tag{1.28}$$

with

$$\operatorname{sdet} \hat{\Delta}^i{}_k = E^{-1\ in} \Delta_{nk} \,. \tag{1.29}$$

The local measure $\mathcal{M}(\varphi)$ can be also chosen in such a way, that the leading ultraviolet divergences in the theory, proportional to the delta-function in coinciding points $\delta(0)$, vanish [115, 117].

1.2 Green Functions of Minimal Differential Operators

The construction of Green functions of arbitrary differential operators (1.22), (1.29) can be reduced finally to the construction of the Green functions of the "minimal" differential operators of second order [35] that have the form

$$\Delta^i{}_k = \{\delta^A{}_B(\Box - m^2) + Q^A{}_B(x)\}\, g^{1/2}(x)\delta(x,x')\,, \tag{1.30}$$

where $\Box = g^{\mu\nu}\nabla_\mu\nabla_\nu$ is the covariant D'Alambert operator, ∇_μ is the covariant derivative, defined by means of some background connection $\mathcal{A}_\mu(x)$,

$$\nabla_\mu\varphi^A = \partial_\mu\varphi^A + \mathcal{A}^A{}_{B\mu}\varphi^B\,, \tag{1.31}$$

$g^{\mu\nu}(x)$ is the metric of the background space-time, $g(x) = -\det g_{\mu\nu}(x)$, m is the mass parameter of the quantum field and $Q^A{}_B(x)$ is an arbitrary matrix-valued function (potential term).

The Green functions $G^A{}_{B'}(x,x')$ of the differential operator (1.30) are two-point objects, which transform as the field $\varphi^A(x)$ under the transformations of coordinates at the point x, and as the current $J_{B'}(x')$ under the coordinate transformations at the point x'. The indices, belonging to the tangent space at the point x', are labeled with a prime.

We will construct solutions of the equation for the Green functions

$$\{\delta^A{}_C(\Box - m^2) + Q^A{}_C\}\, G^C{}_{B'}(x,x') = -\delta^A{}_B g^{-1/2}(x)\delta(x,x')\,, \tag{1.32}$$

with appropriate boundary conditions, by means of the Fock–Schwinger–De Witt proper time method [105, 203, 204, 80, 82] in form of a contour integral over an auxiliary variable s,

$$G = \int\limits_C \mathrm{i}\,ds\, \exp(-ism^2) U(s)\,, \tag{1.33}$$

where the "evolution function" (or the heat kernel) $U(s) \equiv U^A{}_{B'}(s|x,x')$ satisfies the equation

$$\frac{\partial}{\partial is} U(s) = (\hat{1}\,\Box + Q)\, U(s)\,, \qquad \hat{1} \equiv \delta^A{}_B\,, \tag{1.34}$$

with the boundary condition

$$U^A{}_{B'}(s|x,x')\Big|_{\partial C} = -\delta^A{}_B g^{-1/2}(x)\delta(x,x')\,, \tag{1.35}$$

where ∂C is the boundary of the contour C.

The evolution equation (1.34) is as difficult to solve exactly as the initial equation (1.32). However, the representation of the Green functions in form of the contour integrals over the proper time, (1.33), is more convenient to use for the construction of the asymptotic expansion of the Green functions in inverse powers of the mass and for the study of the behavior of the Green functions and their derivatives on the light-cone, $x \to x'$, as well as for the regularization and renormalization of the divergent vacuum expectation values of local variables (such as the energy-momentum tensor, one-loop effective action etc.).

Deforming the contour of integration, C, over s in (1.33) we can get different Green functions for the same evolution function. To obtain the causal Green function (Feynman propagator) one has to integrate over s from 0 to ∞ and add an infinitesimal negative imaginary part to the m^2 [80, 82]. It is this contour that we mean hereafter.

Let us single out in the evolution function a rapidly oscillating factor that reproduces the initial condition (1.35) at $s \to 0$:

$$U(s) = \mathrm{i}(4\pi s)^{-n/2} \Delta^{1/2} \exp\left(-\frac{\sigma}{2\mathrm{i}s}\right) \mathcal{P}\Omega(s) , \tag{1.36}$$

where $\sigma(x, x')$ is half the square of the geodesic distance between the points x and x',

$$\Delta(x, x') = -g^{-1/2}(x) \det\left(-\nabla_{\mu'}\nabla_{\nu}\sigma(x, x')\right) g^{-1/2}(x') \tag{1.37}$$

is the Van Fleck–Morette determinant, $\mathcal{P} \equiv \mathcal{P}^A{}_{B'}(x, x')$ is the parallel displacement operator of the field along the geodesic from the point x' to the point x. We assume that there is only one geodesic connecting the points x and x', the points x and x' being not conjugate, and suppose the two-point functions $\sigma(x, x')$, $\Delta(x, x')$ and $\mathcal{P}^A{}_{B'}(x, x')$ to be single-valued differentiable functions of the coordinates of the points x and x'. When the points x and x' are close enough to each other this will be always the case [211, 80, 82].

The introduced "transfer function" $\Omega(s) \equiv \Omega^{A'}{}_{B'}(s|x, x')$ transforms as a scalar at the point x and as a matrix at the point x' (both its indices are primed). This function is regular in s at the point $s = 0$, i.e.,

$$\Omega^{A'}{}_{B'}(0|x, x')\Big|_{x \to x'} = \delta^{A'}{}_{B'} \tag{1.38}$$

independently on the way how $x \to x'$.

If one assumes that there are no boundary surfaces in space-time (that we will do hereafter), then the is analytic also in x close to the point $x = x'$ for any s, i.e., there exist finite coincidence limits of the and its derivatives at $x = x'$ that do not depend on the way how x approaches x'.

Using the equations for the introduced functions [80, 82],

$$\sigma = \frac{1}{2}\sigma_\mu \sigma^\mu , \qquad \sigma_\mu \equiv \nabla_\mu \sigma , \tag{1.39}$$

$$\sigma^\mu \nabla_\mu \mathcal{P} = 0 , \qquad \mathcal{P}^A{}_{B'}(x', x') = \delta^{A'}{}_{B'} , \tag{1.40}$$

$$\sigma^\mu \nabla_\mu \log \Delta^{1/2} = \frac{1}{2}(n - \Box\sigma) , \tag{1.41}$$

we obtain from (1.34) and (1.36) the transfer equation for the function $\Omega(s)$:

$$\left(\frac{\partial}{\partial is}+\frac{1}{is}\sigma^\mu\nabla_\mu\right)\Omega(s)=\mathcal{P}^{-1}\left(\hat{1}\,\Delta^{-1/2}\,\Box\,\Delta^{1/2}+Q\right)\mathcal{P}\Omega(s)\,. \tag{1.42}$$

If one solves the transfer equation (1.42) in form of a power series in the variable s

$$\Omega(s)=\sum_{k\geq 0}\frac{(is)^k}{k!}a_k\,, \tag{1.43}$$

then from (1.38) and (1.42) one gets the recurrence relations for the a_k

$$\sigma^\mu\nabla_\mu a_0=0\,,\qquad a_0{}^{A'}_{\ B'}(x',x')=\delta^{A'}_{B'}\,, \tag{1.44}$$

$$\left(1+\frac{1}{k}\sigma^\mu\nabla_\mu\right)a_k=\mathcal{P}^{-1}\left(\hat{1}\,\Delta^{-1/2}\,\Box\,\Delta^{1/2}+Q\right)\mathcal{P}a_{k-1}\,. \tag{1.45}$$

The coefficients $a_k(x,x')$ are widely known under the name "heat kernel coefficients", or "HMDS (Hadamard–Minakshisundaram–De Witt–Seeley) coefficients", according to the names of the people who made major contributions to the study of these objects (see [148, 176, 78, 205]). The significance of these coefficients in theoretical and mathematical physics is difficult to overestimate. In this book, following the tradition of the physical literature, we call these coefficients "De Witt coefficients".
From the equations (1.44) it is easy to find the zeroth coefficient

$$a_0{}^{A'}_{\ B'}(x,x')=\delta^{A'}_{B'}\,. \tag{1.46}$$

The other coefficients are calculated usually by differentiating the relations (1.45) and taking the coincidence limits [80, 82]. However such method of calculations is very cumbersome and non-effective. In this way only the coefficients a_1 and a_2 at coinciding points were calculated [80, 62, 63]. The same coefficients as well as the coefficient a_3 at coinciding points were calculated in the paper [132] by means of a completely different non-covariant method. The coefficient a_4 was computed completely independently in [4] and in our papers [12, 11, 9] where a manifestly covariant method for calculation of the De Witt coefficients was elaborated. The coefficient a_5 in the flat space was computed in [219]. Reviews of different methods for computation of the coefficients a_k along with historical comments are presented in [24, 202, 21]. An interesting approach for calculating the heat kernel coefficients was developed in recent papers [188, 189].

In the Chap. 2 we develop a manifestly covariant and very convenient and effective technique that enables to calculate explicitly arbitrary De Witt coefficients a_k as well as to analyze their general structure. This technique was reformulated in [220]. Moreover, the elaborated technique is very algorithmic and can be easily realized on computers [44].

Let us stress that the expansion (1.43) is asymptotic and does not reflect possible nontrivial analytical properties of the transfer function, which are very important when doing the contour integration in (1.33). The expansion in the power series in proper time, (1.43), corresponds physically to the expansion in the dimensionless parameter that is equal to the ratio of the Compton wave length, $\lambda = \hbar/mc$, to the characteristic scale of variation of the background fields, L. That means that it corresponds to the expansion in the Planck constant $\hbar$ in usual units [120, 121]. This is the usual semi-classical approximation of quantum mechanics. This approximation is good enough for the study of the light-cone singularities of the Green functions, for the regularization and renormalization of the divergent coincidence limits of the Green functions and their derivatives at a space-time point, as well as for the calculation of the vacuum polarization of the massive fields in the case when the Compton wave length λ is much smaller than the characteristic length scale L, $\lambda/L = \hbar/(mcL) \ll 1$.

At the same time the expansion in powers of the proper time (1.43) does not contain any information about all effects that depend non-analytically on the Planck constant $\hbar$ (such as the particle creation and the vacuum polarization of massless fields) [82, 121]. Such effects can be described only by summation of the asymptotic expansion (1.43). The exact summation in general case is, obviously, impossible. One can, however, pick up the leading terms in some approximation and sum them up in the first line. Such partial summation of the asymptotic (in general, divergent) series is possible only by employing additional physical assumptions about the analytical structure of the exact expression and corresponding analytical continuation.

In the Chap. 3 we will carry out the partial summation of the terms that are linear and quadratic in background fields as well as the terms without the covariant derivatives of the background fields.

1.3 Divergences, Regularization and Renormalization

A well known problem of quantum field theory is the presence of the ultraviolet divergences that appear in practical calculations in perturbation theory. They are exhibited by the divergence of many integrals (over coordinates and momentums) because of the singular behavior of the Green functions at small distances. The Green functions are, generally speaking, distributions, i.e., linear functionals defined on smooth finite functions [50, 155]. Therefore, numerous products of Green functions appeared in perturbation theory cannot be defined correctly.

A consistent scheme for eliminating the ultraviolet divergences and obtaining finite results is the theory of renormalizations [50, 155], that can be carried out consequently in renormalizable field theories. First of all, one has to introduce an intermediate regularization to give, in some way, the finite values to the formal divergent expressions. Then one should single out the divergent part and include the counter-terms in the classical action that compensate the corresponding divergences. In renormalizable field theories one introduces the counter-terms that have the structure of the individual terms of the classical action. They are interpreted in terms of renormalizations of the fields, the masses and the coupling constants.

By the regularization some new parameters are introduced: a dimensionless regularizing parameter r and a dimensional renormalization parameter μ. After subtracting the divergences and going to the limit $r \to 0$ the regularizing parameter disappears, but the renormalization parameter μ remains and enters the finite renormalized expressions. In renormalized quantum field theories the change of this parameter is compensated by the change of the coupling constants of the renormalized action, $g_i(\mu)$, that are defined at the renormalization point characterized by the energy scale μ. The physical quantities do not depend on the choice of the renormalization point μ where the couplings are defined, i.e., they are renormalization invariant. The transformation of the renormalization parameter μ and the compensating transformations of the parameters of the renormalized action $g_i(\mu)$ form the group of renormalization transformations [50, 226, 229]. The infinitesimal form of these transformations determines the differential equations of the renormalization group that are used for investigating the scaling properties (i.e., the behavior under the homogeneous scale transformation) of the renormalized coupling parameters $g_i(\mu)$, the many-point Green functions and other quantities. In particular, the equations for renormalized coupling constants have the form [229]

$$\mu \frac{d}{d\mu} \bar{g}_i(\mu) = \beta_i(\bar{g}(\mu)) , \tag{1.47}$$

where $\bar{g}_i(\mu) = \mu^{-d_i} g_i(\mu)$ are the dimensionless coupling parameters (d_i is the dimension of the coupling g_i), and $\beta_i(\bar{g})$ are the Gell-Mann–Low β-function.

Let us note, that among the parameters $g_i(\mu)$ there are also non-essential couplings [229] (like the renormalization constants of the fields $Z_r(\mu)$) that are not invariant under the redefinition of the fields. The equations for the renormalization constants $Z_r(\mu)$ have more simple form [229]

$$\mu \frac{d}{d\mu} Z_r(\mu) = \gamma_r(\bar{g}(\mu)) Z_r(\mu) , \tag{1.48}$$

where $\gamma_r(\bar{g})$ are the anomalous dimensions.

The physical quantities (such as the matrix elements of the S-matrix on the mass shell) do not depend on the details of the field definition and, therefore, on the non-essential couplings. On the other hand, the off mass shell Green functions depend on all coupling constants including the non-essential ones. We will apply the renormalization group equations for the investigation of the ultraviolet behavior of the higher-derivative quantum gravity in Chap. 4.

Let us illustrate the procedure of eliminating the ultraviolet divergences by the example of the Green function of the minimal differential operator (1.30) at coinciding points, $G(x,x)$, and the corresponding one-loop effective action $\Gamma_{(1)}$, (1.28). Making use of the Schwinger–De Witt representation for the Green function (1.33) we have

$$G(x,x) = \int_0^\infty \mathrm{i}\,\mathrm{d}s\,\mathrm{i}(4\pi\mathrm{i}s)^{-n/2}\exp(-\mathrm{i}sm^2)\Omega(s|x,x)\ , \qquad (1.49)$$

$$\Gamma_{(1)} = \frac{1}{2}\int_0^\infty \frac{\mathrm{d}s}{s}(4\pi\mathrm{i}s)^{-n/2}\exp(-\mathrm{i}sm^2)\int \mathrm{d}^n x\, g^{1/2}\mathrm{str}\,\Omega(s|x,x)\ . \qquad (1.50)$$

It is clear that in four-dimensional space-time ($n=4$) the integrals over the proper time in (1.49) and (1.50) diverge at the lower limit. Therefore, they should be regularized. To do this one can introduce in the proper time integral a regularizing function $\rho(is\mu^2; r)$ that depends on the regularizing parameter r and the renormalization parameter μ. In the limit $r \to 0$ the regularizing function must tend to unity, and for $r \neq 0$ it must ensure the convergence of the proper time integrals (i.e., it must approach zero sufficiently rapidly at $s \to 0$ and be bounded at $s \to \infty$ by a polynomial). The concrete form of the function ρ does not matter. In practice, one uses the cut-off regularization, the Pauli–Villars one, the analytical one, the dimensional one, the ζ-function regularization and others [42, 50, 155].

The dimensional regularization is one of the most convenient for the practical calculations (especially in massless and gauge theories) as well as for general investigations [207, 193, 42], [60, 212, 163, 54]. The theory is formulated in the space of arbitrary dimension n while the topology and the metric of the additional $n-4$ dimensions can be arbitrary. To preserve the physical dimension of all quantities in the n-dimensional space-time it is necessary to introduce the dimensional parameter μ. All integrals are calculated in that region of the complex plane of n where they converge. It is obvious that for $\mathrm{Re}\, n < C$, with some constant C, the integrals (1.49) and (1.50) converge and define analytic functions of the dimension n. The analytical continuation of these functions to the neighborhood of the physical dimension leads to singularities at the point $n=4$. After subtracting these singularities we obtain analytical functions in the vicinity of the physical dimension, the value of this function at the point $n=4$ defines the finite value of the initial expression.

Let us make some remarks on the dimensional regularization. The analytical continuation of all the relations of the theory to the complex plane of the dimension n is not single-valued, since the values of a function of complex variable at discrete integer values of the argument do not define the unique analytical function [210]. There is also an arbitrariness connected with the subtraction of the divergences. Together with the poles in $(n-4)$ one can also subtract some finite terms (non-minimal renormalization). It is also not necessary to take into account the dependence on the dimension of some quantities (such as the volume element $\mathrm{d}^n x\, g^{1/2}(x)$, background fields, curvatures etc.). On the other hand, one can specify the dependence of all quantities on the additional $n-4$ coordinates in some special explicit way and then calculate the integrals over the $n-4$ dimensions. This would lead to an additional factor that will give, when expanding in $n-4$, additional finite terms. This uncertainty affects only the finite renormalization terms that should be determined from the experiment.

Using the asymptotic expansion (1.43) we obtain in this way from (1.49) and (1.50) the Green function at coinciding points and the one-loop effective action in dimensional regularization

$$\begin{aligned} G(x,x) = \tfrac{\mathrm{i}}{(4\pi)^2}\left\{\left(\tfrac{2}{n-4}+\mathbf{C}+\log\tfrac{m^2}{4\pi\mu^2}\right)(m^2-a_1(x,x))-m^2\right\} \\ +G_{\mathrm{ren}}(x,x)\,, \end{aligned} \tag{1.51}$$

$$G_{\mathrm{ren}}(x,x) = \frac{\mathrm{i}}{(4\pi)^2}\sum_{k\geq 2}\frac{a_k(x,x)}{k(k-1)m^{2(k-1)}}\,, \tag{1.52}$$

$$\begin{aligned} \Gamma_{(1)} = \tfrac{1}{2(4\pi)^2}\Big\{-\tfrac{1}{2}\left(\tfrac{2}{n-4}+\mathbf{C}+\log\tfrac{m^2}{4\pi\mu^2}\right)(m^4-2m^2A_1+A_2) \\ +\tfrac{3}{4}m^4A_0-m^2A_1\Big\}+\Gamma_{(1)\mathrm{ren}}\,, \end{aligned} \tag{1.53}$$

$$\Gamma_{(1)\mathrm{ren}} = \frac{1}{2(4\pi)^2}\sum_{k\geq 3}\frac{A_k}{k(k-1)(k-2)m^{2(k-2)}}\,, \tag{1.54}$$

where $\mathbf{C}\approx 0.577$ is the Euler constant and

$$A_k \equiv \int \mathrm{d}^n x\, g^{1/2}\mathrm{str}\, a_k(x,x)\,. \tag{1.55}$$

Here all the coefficients a_k and A_k are n-dimensional. However in that part, which is analytical in $n-4$, one can treat them as 4-dimensional.

2. Technique for Calculation of De Witt Coefficients

2.1 Covariant Expansions in Curved Space

Let us single out a small regular region in the space, fix a point x' in it, and connect any other point x with the point x' by a geodesic $x = x(\tau)$, $x(0) = x'$, where τ is an affine parameter.

The world function $\sigma(x, x')$ (in terminology of [211]), introduced in the Chap. 1, has the form

$$\sigma(x, x') = \frac{1}{2}\tau^2 \dot{x}^2(\tau) , \tag{2.1}$$

where

$$\dot{x}^\mu(\tau) = \frac{d}{d\tau} x^\mu(\tau) . \tag{2.2}$$

The first derivatives of the function $\sigma(x, x')$ with respect to coordinates are proportional to the tangent vectors to the geodesic at the points x and x' [211],

$$\sigma^\mu = \tau \dot{x}^\mu(\tau) , \qquad \sigma^{\mu'} = -\tau \dot{x}^\mu(0) , \tag{2.3}$$

where

$$\sigma_\mu = \nabla_\mu \sigma , \qquad \sigma_{\mu'} = \nabla_{\mu'} \sigma .$$

From here, it follows, in particular, the basic identity (1.39), that the function $\sigma(x, x')$ satisfies,

$$(D - 2)\sigma = 0 , \qquad D \equiv \sigma^\mu \nabla_\mu , \tag{2.4}$$

the coincidence limits

$$[\sigma] = [\sigma^\mu] = [\sigma^{\mu'}] = 0 , \tag{2.5}$$

$$[f(x, x')] \equiv \lim_{x \to x'} f(x, x') , \tag{2.6}$$

and the relation between the tangent vectors

$$\sigma^\mu = -g^\mu{}_{\nu'} \sigma^{\nu'} , \tag{2.7}$$

where $g^\mu{}_{\nu'}(x, x')$ is the parallel displacement operator of vectors along the geodesic from the point x' to the point x. The non-primed (primed) indices are lowered and risen by the metric tensor in the point x (x').

By differentiating the basic identity (2.4) we obtain the relations

$$(D-1)\sigma^{\mu}=0\,,\qquad \sigma^{\mu}=\xi^{\mu}{}_{\nu}\sigma^{\nu}\,, \tag{2.8}$$

$$(D-1)\sigma^{\mu'}=0,\qquad \sigma^{\mu'}=\eta^{\mu'}{}_{\nu}\sigma^{\nu}\,, \tag{2.9}$$

where

$$\xi^{\mu}{}_{\nu}=\nabla_{\nu}\sigma^{\mu}\,,\qquad \eta^{\mu'}{}_{\nu}=\nabla_{\nu}\sigma^{\mu'}\,. \tag{2.10}$$

Therefrom the coincidence limits follow

$$[\xi^{\mu}{}_{\nu}]=-[\eta^{\mu'}{}_{\nu}]=\delta^{\mu}{}_{\nu}\,, \tag{2.11}$$

$$[\nabla_{(\mu_1}\cdots\nabla_{\mu_k)}\sigma^{\nu}]=[\nabla_{(\mu_1}\cdots\nabla_{\mu_k)}\sigma^{\nu'}]=0\,,\qquad (k\geq 2)\,. \tag{2.12}$$

Let us consider a field $\varphi=\varphi^A(x)$ and an affine connection $\mathcal{A}_\mu=\mathcal{A}^A{}_{B\mu}(x)$, that defines the covariant derivative (1.31) and the commutator of covariant derivatives,

$$[\nabla_\mu,\nabla_\nu]\varphi=\mathcal{R}_{\mu\nu}\varphi\,, \tag{2.13}$$

$$\mathcal{R}_{\mu\nu}=\partial_\mu\mathcal{A}_\nu-\partial_\nu\mathcal{A}_\mu+[\mathcal{A}_\mu,\mathcal{A}_\nu]\,. \tag{2.14}$$

Let us define the parallel displacement operator of the field φ along the geodesic from the point x' to the point x, $\mathcal{P}=\mathcal{P}^A{}_{B'}(x,x')$, to be the solution of the equation of parallel transport,

$$D\mathcal{P}=0\,, \tag{2.15}$$

with the initial condition

$$[\mathcal{P}]=\mathcal{P}(x,x)=\hat{1}\,. \tag{2.16}$$

From here one can obtain the coincidence limits

$$\left[\nabla_{(\mu_1}\cdots\nabla_{\mu_k)}\mathcal{P}\right]=0\,,\qquad (k\geq 1)\,. \tag{2.17}$$

In particular, when $\varphi=\varphi^\mu$ is a vector field, and the connection $\mathcal{A}_\mu=\Gamma^\alpha{}_{\mu\beta}$ is the Christoffel connection, the equations (2.15) and (2.16) define the parallel displacement operator of the vectors: $\mathcal{P}=g^{\mu}{}_{\nu'}(x,x')$.

Let us transport the field φ parallel along the geodesic to the point x'

$$\bar\varphi=\bar\varphi^{C'}(x)=\mathcal{P}^{C'}_{A}(x',x)\varphi^A(x)=\mathcal{P}^{-1}\varphi\,, \tag{2.18}$$

where $\mathcal{P}^{-1}=\mathcal{P}^{C'}_{A}(x',x)$ is the parallel displacement operator along the opposite path (from the point x to the point x' along the geodesic):

$$\mathcal{P}\mathcal{P}^{-1}=\hat{1}\,. \tag{2.19}$$

The obtained object $\bar\varphi$, (2.18), is a scalar under the coordinate transformations at the point x, since it does not have any non-prime indices. By considering $\bar\varphi$ as a function of the affine parameter τ, let us expand it in the Taylor series

$$\bar{\varphi} = \sum_{k\geq 0} \frac{1}{k!}\tau^k \left[\frac{d^k}{d\tau^k}\bar{\varphi}\right]_{\tau=0} . \tag{2.20}$$

Noting that $d/d\tau = \dot{x}^\mu \partial_\mu$, $\partial_\mu \bar{\varphi} = \nabla_\mu \bar{\varphi}$ and using the equation of the geodesic, $\dot{x}^\mu \nabla_\mu \dot{x}^\nu = 0$, and the equations (2.3), (2.9) and (2.18), we obtain

$$\varphi = \mathcal{P} \sum_{k\geq 0} \frac{(-1)^k}{k!} \sigma^{\mu_1'} \cdots \sigma^{\mu_k'} \varphi_{\mu_1' \cdots \mu_k'} , \tag{2.21}$$

where

$$\varphi_{\mu_1' \cdots \mu_k'} = \left[\nabla_{(\mu_1} \cdots \nabla_{\mu_k)} \varphi\right] . \tag{2.22}$$

The equation (2.21) is the generalized covariant Taylor series for arbitrary field with arbitrary affine connection in a curved space.

Let us show that the series (2.21) is the expansion in a complete set of eigenfunctions of the operator D, (2.4). The vectors σ^μ and $\sigma^{\mu'}$ are the eigenfunctions of the operator D with the eigenvalues equal to 1 (see (2.8) and (2.9)). Therefore, one can construct the eigenfunctions with arbitrary positive integer eigenvalues:

$$|0>\equiv 1,$$

$$|n>\equiv |\nu_1' ... \nu_n'> = \frac{(-1)^n}{n!} \sigma^{\nu_1'} \cdots \sigma^{\nu_n'} , \qquad (n \geq 1) , \tag{2.23}$$

$$D|n> = n|n> . \tag{2.24}$$

We have

$$|n_1> \otimes \cdots \otimes |n_k> = \binom{n}{n_1, \ldots, n_k} |n> , \tag{2.25}$$

where

$$\binom{n}{n_1, \ldots, n_k} = \frac{n!}{n_1! \cdots n_k!} , \qquad n = n_1 + \cdots + n_k . \tag{2.26}$$

Let us note that there exist more general eigenfunctions of the form $(\sigma)^z |n>$ with arbitrary eigenvalues $(n + 2z)$

$$D\,(\sigma)^z |n> = (n + 2z)(\sigma)^z |n> . \tag{2.27}$$

However, for non-integer or negative z these functions are not analytic in coordinates of the point x in the vicinity of the point x'. For positive integer z they reduce to the linear combinations of the functions (2.23). Therefore, we restrict ourselves to the functions (2.23) having in mind to study only regular fields near the point x'.

Let us introduce the dual functions

$$< m| \equiv < \mu_1' \cdots \mu_m'| = (-1)^m g^{\mu_1}_{\mu_1'} \cdots g^{\mu_m}_{\mu_m'} \nabla_{(\mu_1} \cdots \nabla_{\mu_m)} \delta(x, x') \tag{2.28}$$

and the scalar product

$$< m|n >= \int \mathrm{d}^n x \, < \mu_1' \cdots \mu_m' | \nu_1' ... \nu_n' > \, . \tag{2.29}$$

Using the coincidence limits (2.11) and (2.12) it is easy to prove that the set of the eigenfunctions (2.23) and (2.28) is orthonormal

$$< m|n >= \delta_{mn} 1_{(n)} \,, \qquad 1_{(n)} \equiv \delta^{\nu_1 \cdots \nu_n}_{\mu_1 \cdots \mu_n} = \delta^{\nu_1}_{(\mu_1} \cdots \delta^{\nu_n}_{\mu_n)} \, . \tag{2.30}$$

The introduced scalar product (2.29) reduces to the coincidence limit of the symmetrized covariant derivatives,

$$< m|\varphi >= \left[\nabla_{(\mu_1} \cdots \nabla_{\mu_m)} \varphi \right] \, . \tag{2.31}$$

Therefore, the covariant Taylor series (2.21) can be rewritten in a compact form

$$|\varphi >= \mathcal{P} \sum_{n \geq 0} |n >< n|\varphi > \, . \tag{2.32}$$

From here it follows the condition of completeness of the set of the eigenfunctions (2.23) in the space of scalar functions regular in the vicinity of the point x'

$$\mathbf{I} = \sum_{n \geq 0} |n >< n| \,, \tag{2.33}$$

or, more precisely,

$$\delta(x, y) = \sum_{n \geq 0} \frac{1}{n!} \sigma^{\mu_1'}(x, x') \cdots \sigma^{\mu_n'}(x, x') g^{\mu_1}_{\mu_1'}(y, x') \cdots g^{\mu_n}_{\mu_n'}(y, x')$$

$$\times \nabla^y_{(\mu_1} \cdots \nabla^y_{\mu_n)} \delta(y, x') \, . \tag{2.34}$$

Let us note that, since the parallel displacement operator $\mathcal{P}$ is an eigenfunction of the operator D with zero eigenvalue, (2.15), one can also introduce a complete orthonormal set of "isotopic" eigenfunctions $\mathcal{P}|n>$ and $< n|\mathcal{P}^{-1}$.

The complete set of eigenfunctions (2.23) can be employed to present an arbitrary linear differential operator F defined on the fields φ in the form

$$F = \sum_{m,n \geq 0} \mathcal{P}|m >< m|\mathcal{P}^{-1} F \mathcal{P}|n >< n|\mathcal{P}^{-1} \,, \tag{2.35}$$

where

$$< m|\mathcal{P}^{-1} F \mathcal{P}|n >= \left[\nabla_{(\mu_1} \cdots \nabla_{\mu_m)} \mathcal{P}^{-1} F \mathcal{P} \frac{(-1)^n}{n!} \sigma^{\nu_1'} \cdots \sigma^{\nu_n'} \right] \tag{2.36}$$

are the "matrix elements" of the operator F (2.35). The matrix elements (2.36) are expressed finally in terms of the coincidence limits of the derivatives of the coefficient functions of the operator F, the parallel displacement operator $\mathcal{P}$ and the world function σ.

For calculation of the matrix elements of differential operators (2.36) as well as for constructing the covariant Fourier integral it is convenient to make a change of the variables

$$x^\mu = x^\mu(\sigma^{\nu'}, x^{\lambda'}) , \qquad (2.37)$$

i.e., to consider a function of the coordinates x^μ as the function of the vectors $\sigma^{\nu'}(x, x')$ and the coordinates $x^{\lambda'}$.

The derivatives and the differentials in old and new variables are connected by the relations

$$\partial_\mu = \eta^{\nu'}{}_\mu \bar{\partial}_{\nu'} , \qquad \bar{\partial}_{\nu'} = \gamma^\mu{}_{\nu'} \partial_\mu ,$$

$$dx^\mu = \gamma^\mu{}_{\nu'} d\sigma^{\nu'} , \qquad d\sigma^{\nu'} = \eta^{\nu'}{}_\mu dx^\mu , \qquad (2.38)$$

where $\partial_\mu = \partial/\partial x^\mu$, $\bar{\partial}_{\mu'} = \partial/\partial\sigma^{\mu'}$, $\eta^{\nu'}{}_\mu$ is defined in (2.10), $\gamma^\mu{}_{\nu'}$ are the elements of the inverse matrix,

$$\gamma = \eta^{-1} , \qquad (2.39)$$

and η is a matrix with elements $\eta^{\nu'}{}_\mu$.

From the coincidence limits (2.11) it follows that for close points x and x'

$$\det \eta \neq 0 , \qquad \det \gamma \neq 0 , \qquad (2.40)$$

and, therefore, the change of variables (2.37) is admissible.

The corresponding covariant derivatives are connected by analogous relations

$$\nabla_\mu = \eta^{\nu'}{}_\mu \bar{\nabla}_{\nu'} , \qquad \bar{\nabla}_{\nu'} = \gamma^\mu{}_{\nu'} \nabla_\mu . \qquad (2.41)$$

From the definition of the matrices η, (2.10), and γ, (2.39), one can get the relations

$$\nabla_{[\lambda} \eta^{\mu'}{}_{\nu]} = 0 , \qquad \bar{\nabla}_{[\lambda'} \gamma^\mu{}_{\nu']} = 0 , \qquad (2.42)$$

$$\bar{\nabla}_{\mu'} \left(\Delta^{-1} \eta^{\mu'}{}_\nu \right) = 0 , \qquad (2.43)$$

where Δ is the Van Fleck–Morette determinant (1.37)

$$\begin{aligned} \Delta(x, x') &= g^{1/2}(x') g^{-1/2}(x) \det(-\eta) = g^{1/2}(x') g^{-1/2}(x) \det(-\gamma)^{-1} \\ &= \det(-\bar{\eta}) = (\det(-\bar{\gamma}))^{-1} = (\det X)^{1/2} , \end{aligned} \qquad (2.44)$$

and $\bar{\eta}$, $\bar{\gamma}$ and X are matrices with elements

$$\bar{\eta}^{\mu'}{}_{\nu'} = g^\nu{}_{\nu'} \eta^{\mu'}{}_\nu , \qquad \bar{\gamma}^{\mu'}{}_{\nu'} = g^{\mu'}{}_\mu \gamma^\mu{}_{\nu'} , \qquad (2.45)$$

$$X^{\mu'\nu'} = \eta^{\mu'}{}_\mu g^{\mu\nu} \eta^{\nu'}{}_\nu . \qquad (2.46)$$

Let us note that the dual eigenfunctions (2.28) of the operator D can be expressed in terms of the operators $\bar\nabla$, (2.41),

$$< m| \equiv < \mu'_1 \cdots \mu'_m| = \bar\nabla_{(\mu'_1} \cdots \bar\nabla_{\mu'_m)}\delta(x,x') . \tag{2.47}$$

Therefore, the coefficients of the covariant Taylor series (2.21), (2.22), (2.31) and (2.32) can be also written in terms of the operators $\bar\nabla$:

$$< m|\varphi >= (-1)^m \left[\bar\nabla_{(\mu'_1} \cdots \bar\nabla_{\mu'_m)}\varphi\right] . \tag{2.48}$$

The commutator of the operators $\bar\nabla$, when acting on the parallel displacement operator $\mathcal{P}$, has the form

$$[\bar\nabla_{\mu'}, \bar\nabla_{\nu'}]\mathcal{P} = \bar{\mathcal{R}}_{\mu'\nu'}\mathcal{P} , \tag{2.49}$$

where

$$\begin{aligned} \bar{\mathcal{R}}_{\mu'\nu'} &= \gamma^\mu{}_{\mu'}\gamma^\nu{}_{\nu'}\mathcal{P}^{-1}\mathcal{R}_{\mu\nu}\mathcal{P} \\ &= \bar\nabla_{\mu'}\bar{\mathcal{A}}_{\nu'} - \bar\nabla_{\nu'}\bar{\mathcal{A}}_{\mu'} + [\bar{\mathcal{A}}_{\mu'}, \bar{\mathcal{A}}_{\nu'}] , \end{aligned} \tag{2.50}$$

$$\bar{\mathcal{A}}_{\mu'} = \mathcal{P}^{-1}\bar\nabla_{\mu'}\mathcal{P} . \tag{2.51}$$

The quantity (2.51) introduced here satisfies the equation (2.15),

$$\sigma^{\mu'}\bar{\mathcal{A}}_{\mu'} = 0 . \tag{2.52}$$

On the other hand, when acting on the objects $\bar\varphi$, (2.18), that do not have non-primed indices, the operators $\bar\nabla$ commute with each other

$$[\bar\nabla_{\mu'}, \bar\nabla_{\nu'}]\bar\varphi = 0 . \tag{2.53}$$

Thus the vectors $\sigma^{\nu'}$ and the operators $\bar\nabla_{\mu'}$, (2.41), play the role analogous to that of usual coordinates and the operator of differentiation in the tangent space at the point x'. In particular,

$$[\bar\nabla_{\mu'}, \sigma^{\nu'}] = \delta^{\nu'}{}_{\mu'} . \tag{2.54}$$

Therefore, one can construct the covariant Fourier integral in the tangent space at the point x' in the usual way using the variables $\sigma^{\mu'}$. So that for the fields $\bar\varphi$, (2.18), we have

$$\begin{aligned} \bar\varphi(k) &= \int \mathrm{d}^n x\, g^{1/2}(x)\Delta(x,x')\exp(\mathrm{i}k_{\mu'}\sigma^{\mu'})\bar\varphi(x) , \\ \bar\varphi(x) &= \int \frac{\mathrm{d}^n k^{\mu'}}{(2\pi)^n} g^{1/2}(x')\exp(-\mathrm{i}k_{\mu'}\sigma^{\mu'})\bar\varphi(k) . \end{aligned} \tag{2.55}$$

Note, that the standard rule,

$$\bar{\nabla}_{\mu'}\bar{\varphi}(x) = \int \frac{\mathrm{d}^n k^{\mu'}}{(2\pi)^n} g^{1/2}(x') \exp(-\mathrm{i}k_{\mu'}\sigma^{\mu'})(-\mathrm{i}k_{\mu'})\bar{\varphi}(k) , \tag{2.56}$$

takes place and the covariant momentum representation of the delta-function has the form

$$\delta(x,y) = \int \frac{\mathrm{d}^n k^{\mu'}}{(2\pi)^n} g^{1/2}(x') g^{1/2}(x) \Delta(x,x')$$

$$\times \exp\left\{\mathrm{i}k_{\mu'}\left(\sigma^{\mu'}(y,x') - \sigma^{\mu'}(x,x')\right)\right\} . \tag{2.57}$$

2.2 Elements of Covariant Expansions

Let us calculate the quantities η, γ and X introduced in previous section. By differentiating the equations (2.8) and (2.9) and commuting the covariant derivatives we get

$$D\xi + \xi(\xi - \hat{1}) + S = 0 , \tag{2.58}$$

$$D\eta + \eta(\xi - \hat{1}) = 0 , \tag{2.59}$$

where

$$\xi = \xi^{\mu}{}_{\nu} , \qquad S = S^{\mu}{}_{\nu} , \qquad \hat{1} = \delta^{\mu}{}_{\nu} ,$$

$$S^{\mu}{}_{\nu} = R^{\mu}{}_{\alpha\nu\beta}\sigma^{\alpha}\sigma^{\beta} . \tag{2.60}$$

By solving (2.59) with respect to the matrix ξ, substituting the solution in (2.58) and taking into account (2.39) we obtain the linear equation for the matrix $\bar{\gamma}$ (2.45)

$$\{\hat{1}\,(D^2 + D) + \bar{S}\}\,\bar{\gamma} = 0 , \tag{2.61}$$

where $\bar{S} = \bar{S}^{\mu'}{}_{\nu'} = g^{\mu'}{}_{\mu} g^{\nu}{}_{\nu'} S^{\mu}{}_{\nu}$, with the boundary condition, (2.39), (2.11),

$$[\bar{\gamma}] = -1 . \tag{2.62}$$

One can solve the equation (2.61) perturbatively, treating the matrix $\bar{S}$ as a perturbation. Supposing

$$\bar{\gamma} = -\hat{1} + \hat{\bar{\gamma}} , \tag{2.63}$$

we obtain

$$\{\hat{1}\,(D^2 + D) + \bar{S}\}\,\hat{\bar{\gamma}} = \bar{S} , \qquad [\hat{\bar{\gamma}}] = 0 . \tag{2.64}$$

From here we have formally

$$\hat{\bar{\gamma}} = \{\hat{1}\,(D^2 + D) + \bar{S}\}^{-1}\bar{S} = \sum_{k\geq 1}(-1)^{k+1}\left\{(D^2 + D)^{-1}\bar{S}\right\}^k \cdot \hat{1} . \tag{2.65}$$

The formal expression (2.65) becomes meaningful in terms of the expansion in the eigenfunctions of the operator D, (2.23). The inverse operator

$$(D^2 + D)^{-1} = \sum_{n \geq 0} \frac{1}{n(n+1)} |n><n| \, , \tag{2.66}$$

when acting on the matrix $\bar{S}$, is well defined, since $<0|\bar{S}>=0$. Expanding the matrix $\bar{S}$ in the covariant Taylor matrix according to (2.21) and (2.32),

$$\bar{S} = \sum_{n \geq 2} \frac{(-1)^n}{(n-2)!} K_{(n)} \, , \tag{2.67}$$

where $K_{(n)}$ is the matrix with entries $K^{\mu'}{}_{\nu'(n)}$

$$K^{\mu'}{}_{\nu'(n)} = K^{\mu'}{}_{\nu'\mu'_1 \cdots \mu'_n} \sigma^{\mu'_1} \cdots \sigma^{\mu'_n} \, ,$$

$$K^{\mu}{}_{\nu\mu_1 \cdots \mu_n} = \nabla_{(\mu_1} \cdots \nabla_{\mu_{n-2}} R^{\mu}{}_{\mu_{n-1}|\nu|\mu_n)} \, , \tag{2.68}$$

we obtain

$$\bar{\gamma} = -\hat{1} + \sum_{n \geq 2} \frac{(-1)^n}{n!} \gamma_{(n)} \, , \tag{2.69}$$

$$\begin{aligned} \gamma_{(n)} &= \gamma_{\mu'_1 \cdots \mu'_n} \sigma^{\mu'_1} \cdots \sigma^{\mu'_n} \\ &= \sum_{1 \leq k \leq [n/2]} (-1)^{k+1} (2k)! \binom{n}{2k} \sum_{\substack{n_1, \cdots, n_k \geq 2 \\ n_1 + \cdots + n_k = n}} \binom{n-2k}{n_1 - 2, \cdots, n_k - 2} \\ &\quad \times \frac{K_{(n_k)}}{n(n+1)} \cdot \frac{K_{(n_k - 1)}}{(n_1 + \cdots + n_{k-1})(n_1 + \cdots + n_{k-1} + 1)} \\ &\quad \times \cdots \frac{K_{(n_2)}}{(n_1 + n_2)(n_1 + n_2 + 1)} \cdot \frac{K_{(n_1)}}{n_1(n_1 + 1)} \, , \end{aligned} \tag{2.70}$$

where

$$\binom{n}{n_1, \cdots, n_k} = \frac{n!}{n_1! \cdots n_k!} \, ,$$

with $n = n_1 + \cdots + n_k$ for natural numbers $n_1, \ldots, n_k$.

Let us write down some first coefficients (2.70):

$$\begin{aligned} \gamma^{\alpha}{}_{\beta\mu_1\mu_2} &= \frac{1}{3} R^{\alpha}{}_{(\mu_1|\beta|\mu_2)} \, , \\ \gamma^{\alpha}{}_{\beta\mu_1\mu_2\mu_3} &= \frac{1}{2} \nabla_{(\mu_1} R^{\alpha}{}_{\mu_2|\beta|\mu_3)} \, , \end{aligned} \tag{2.71}$$

$$\gamma^{\alpha}{}_{\beta\mu_1\mu_2\mu_3\mu_4} = \frac{3}{5}\nabla_{(\mu_1}\nabla_{\mu_2}R^{\alpha}{}_{\mu_3|\beta|\mu_4)} - \frac{1}{5}R^{\alpha}{}_{(\mu_1|\gamma|\mu_2}R^{\gamma}{}_{\mu_3|\beta|\mu_4)} \,.$$

Using the solution (2.69) one can find all other quantities. The inverse matrix $\bar\eta = \bar\gamma^{-1}$ has the form

$$\bar\eta = -\hat 1 + \sum_{n\geq 2}\frac{(-1)^n}{n!}\eta_{(n)} \,, \tag{2.72}$$

$$\begin{aligned}\eta_{(n)} &= \eta_{\mu'_1\cdots\mu'_n}\sigma^{\mu'_1}\cdots\sigma^{\mu'_n} \\ &= -\sum_{1\leq k\leq [n/2]}\ \sum_{\substack{n_1,\cdots,n_k\geq 2\\ n_1+\cdots+n_k=n}}\binom{n}{n_1,\cdots n_k}\gamma_{(n_k)}\cdots\gamma_{(n_1)} \,.\end{aligned} \tag{2.73}$$

Some first coefficients (2.73) equal

$$\begin{aligned}\eta^{\alpha}{}_{\beta\mu_1\mu_2} &= -\frac{1}{3}R^{\alpha}{}_{(\mu_1|\beta|\mu_2)} \,, \\ \eta^{\alpha}{}_{\beta\mu_1\mu_2\mu_3} &= -\frac{1}{2}\nabla_{(\mu_1}R^{\alpha}{}_{\mu_2|\beta|\mu_3)} \,, \\ \eta^{\alpha}{}_{\beta\mu_1\mu_2\mu_3\mu_4} &= -\frac{3}{5}\nabla_{(\mu_1}\nabla_{\mu_2}R^{\alpha}{}_{\mu_3|\beta|\mu_4)} - \frac{7}{15}R^{\alpha}{}_{(\mu_1|\gamma|\mu_2}R^{\gamma}{}_{\mu_3|\beta|\mu_4)} \,.\end{aligned} \tag{2.74}$$

Using (2.64) and (2.72) we find the matrix X:

$$X^{\mu'\nu'} = g^{\mu'\nu'} + \sum_{n\geq 2}\frac{(-1)^n}{n!}X^{\mu'\nu'}{}_{(n)} \,, \tag{2.75}$$

$$\begin{aligned}X^{\mu'\nu'}{}_{(n)} &= X^{\mu'\nu'}{}_{\mu'_1\cdots\mu'_n}\sigma^{\mu'_1}\cdots\sigma^{\mu'_n} \\ &= -2\eta^{(\mu'\nu')}{}_{(n)} + \sum_{2\leq k\leq n-2}\binom{n}{k}\eta^{(\mu'}{}_{\alpha(n-k)}\eta^{\nu')\alpha}{}_{(k)} \,.\end{aligned} \tag{2.76}$$

The lowest order coefficients (2.76) read

$$\begin{aligned}X^{\mu\nu}{}_{\mu_1\mu_2} &= \frac{2}{3}R^{\mu}{}_{(\mu_1}{}^{\nu}{}_{\mu_2)} \,, \\ X^{\mu\nu}{}_{\mu_1\mu_2\mu_3} &= \nabla_{(\mu_1}R^{\mu}{}_{\mu_2}{}^{\nu}{}_{\mu_3)} \,, \\ X^{\mu\nu}{}_{\mu_1\mu_2\mu_3\mu_4} &= \frac{6}{5}\nabla_{(\mu_1}\nabla_{\mu_2}R^{\mu}{}_{\mu_3}{}^{\nu}{}_{\mu_4)} + \frac{8}{5}R^{\mu}{}_{(\mu_1|\alpha|\mu_2}R^{\alpha}{}_{\mu_3}{}^{\nu}{}_{\mu_4)} \,.\end{aligned} \tag{2.77}$$

Finally, we find the Van Fleck–Morette determinant (2.44):

$$\Delta = \exp(2\zeta) \,,$$

$$\zeta = \sum_{n\ge 2} \frac{(-1)^n}{n!} \zeta_{(n)} \,, \tag{2.78}$$

$$\begin{aligned}
\zeta_{(n)} &= \zeta_{\mu'_1\cdots\mu'_n}\sigma^{\mu'_1}\cdots\sigma^{\mu'_n} \\
&= \sum_{1\le k\le [n/2]} \frac{1}{2k} \sum_{\substack{n_1,\cdots,n_k\ge 2\\ n_1+\cdots+n_k=n}} \binom{n}{n_1,\cdots,n_k} \mathrm{tr}\left(\gamma_{(n_1)}\cdots\gamma_{(n_k)}\right) \,.
\end{aligned} \tag{2.79}$$

The first coefficients (2.79) equal

$$\begin{aligned}
\zeta_{\mu_1\mu_2} &= \frac{1}{6} R_{\mu_1\mu_2} \,, \\
\zeta_{\mu_1\mu_2\mu_3} &= \frac{1}{4} \nabla_{(\mu_1} R_{\mu_2\mu_3)} \,, \\
\zeta_{\mu_1\mu_2\mu_3\mu_4} &= \frac{3}{10} \nabla_{(\mu_1}\nabla_{\mu_2} R_{\mu_3\mu_4)} + \frac{1}{15} R_{\alpha(\mu_1|\gamma|\mu_2} R^{\gamma}{}_{\mu_3}{}^{\alpha}{}_{\mu_4)} \,.
\end{aligned} \tag{2.80}$$

Let us calculate the quantity $\bar{\mathcal{A}}_{\mu'}$, (2.51). By differentiating the equation (2.52) and using (2.50) we obtain

$$(D+1)\bar{\mathcal{A}}_{\mu'} = -\bar{\gamma}^{\nu'}{}_{\mu'}\bar{\mathcal{L}}_{\nu'} \,, \tag{2.81}$$

where

$$\bar{\mathcal{L}}_{\nu'} = g^{\mu}{}_{\nu'}\mathcal{P}^{-1}\mathcal{R}_{\mu\alpha}\mathcal{P}\sigma^{\alpha} \,. \tag{2.82}$$

From here we have

$$\bar{\mathcal{A}}_{\mu'} = -(D+1)^{-1}\bar{\gamma}^{\nu'}{}_{\mu'}\bar{\mathcal{L}}_{\nu'} \,, \tag{2.83}$$

where the inverse operator $(D+1)^{-1}$ is defined by

$$(D+1)^{-1} = \sum_{n\ge 0} \frac{1}{n+1} |n><n| \,.$$

Expanding the vector $\bar{\mathcal{L}}_{\mu'}$, (2.82), in the covariant Taylor series (2.21), (2.32)

$$\bar{\mathcal{L}}_{\mu'} = \sum_{n\ge 1} \frac{(-1)^n}{(n-1)!} \mathcal{R}_{\mu'(n)} \,, \tag{2.85}$$

where

$$\begin{aligned}
\mathcal{R}_{\mu'(n)} &= \mathcal{R}_{\mu'\mu'_1\cdots\mu'_n}\sigma^{\mu'_1}\cdots\sigma^{\mu'_n} \,, \\
\mathcal{R}^{\mu}{}_{\mu_1\cdots\mu_n} &= \nabla_{(\mu_1}\cdots\nabla_{\mu_{n-1}}\mathcal{R}^{\mu}{}_{\mu_n)} \,,
\end{aligned} \tag{2.86}$$

we obtain

$$\bar{\mathcal{A}}_{\mu'} = \sum_{n\geq 1} \frac{(-1)^n}{n!} \mathcal{A}_{\mu'(n)} \,, \tag{2.87}$$

$$\begin{aligned} \mathcal{A}_{\mu'(n)} &= \mathcal{A}_{\mu'\mu'_1\cdots\mu'_n} \sigma^{\mu'_1} \cdots \sigma^{\mu'_n} \\ &= \frac{n}{n+1} \left\{ \mathcal{R}_{\mu'(n)} - \sum_{2\leq k\leq n-1} \binom{n-1}{k} \gamma^{\alpha'}_{\ \mu'(k)} \mathcal{R}_{\alpha'(n-k)} \right\} . \end{aligned} \tag{2.88}$$

In particular, the first coefficients (2.88) have the form

$$\begin{aligned} \mathcal{A}_{\mu\mu_1} &= \frac{1}{2} \mathcal{R}_{\mu\mu_1} \,, \\ \mathcal{A}_{\mu\mu_1\mu_2} &= \frac{2}{3} \nabla_{(\mu_1} \mathcal{R}_{|\mu|\mu_2)} \,, \\ \mathcal{A}_{\mu\mu_1\mu_2\mu_3} &= \frac{3}{4} \nabla_{(\mu_1} \nabla_{\mu_2} \mathcal{R}_{|\mu|\mu_3)} - \frac{1}{4} R_{\alpha(\mu_1|\mu|\mu_2} \mathcal{R}^{\alpha}_{\ \mu_3)} \,. \end{aligned} \tag{2.89}$$

Thus we have obtained all the needed quantities in form of covariant Taylor series: (2.69), (2.72), (2.75), (2.78) and (2.87).

In particular case, when all the derivatives of the curvature tensors can be neglected,

$$\nabla_\mu R^\alpha_{\ \beta\gamma\delta} = 0 \,, \qquad \nabla_\mu \mathcal{R}_{\alpha\beta} = 0 \,, \tag{2.90}$$

one can solve exactly the equations (2.61), (2.81) or, equivalently, sum up all terms in the Taylor series, which do not contain the derivatives of the curvature tensors. In this case the Taylor series (2.69) is a power series in the matrix $\bar{S}$. It is an eigenmatrix of the operator D,

$$D\bar{S} = 2\bar{S}, \tag{2.91}$$

and, therefore, when acting on the series (2.69), one can present the operator D in the form

$$D = 2\bar{S} \frac{d}{d\bar{S}} \,, \tag{2.92}$$

and treat the matrix $\bar{S}$ as an usual scalar variable. Substituting (2.92) in (2.61), we obtain an ordinary second order differential equation

$$\left(\frac{d^2}{dt^2} + \frac{2}{t}\frac{d}{dt} + \hat{1} \right) \bar{\gamma} = 0 \,, \qquad t \equiv \sqrt{\bar{S}} \,, \tag{2.93}$$

that has the general solution

$$\bar{\gamma} = (\bar{S})^{-1/2} \left(C_1 \sin\sqrt{\bar{S}} + C_2 \cos\sqrt{\bar{S}} \right) \,, \tag{2.94}$$

where C_1 and C_2 are the integration constants. Using the initial condition (2.62), we have finally $C_1 = -1$, $C_2 = 0$, i.e.,

$$\bar{\gamma} = -\frac{\sin\sqrt{\bar{S}}}{\sqrt{\bar{S}}} . \tag{2.95}$$

In any case this formal expression makes sense as the series (2.69). It is certain to converge for close points x and x'.

Using (2.95), (2.39) and (2.46) we find easily the matrices $\bar{\eta}$ and X

$$\bar{\eta} = -\frac{\sqrt{\bar{S}}}{\sin\sqrt{\bar{S}}} ,$$

$$X = \frac{\bar{S}}{\sin^2\sqrt{\bar{S}}} , \tag{2.96}$$

and the Van Fleck–Morette determinant Δ, (2.44),

$$\Delta = \det\left(\frac{\sqrt{\bar{S}}}{\sin\sqrt{\bar{S}}}\right) . \tag{2.97}$$

The vector $\bar{\mathcal{A}}_{\mu'}$, (2.87), in the case (2.90) is presented as the product of some matrix $H^{\mu'}{}_{\nu'}$, that does not depend on $\mathcal{R}_{\mu\nu}$, and the vector $\bar{\mathcal{L}}_{\mu'}$, (2.82),

$$\bar{\mathcal{A}}_{\mu'} = H_{\mu'\nu'}\bar{\mathcal{L}}^{\nu'} . \tag{2.98}$$

Substituting (2.98) in (2.81) and using the equation

$$D\bar{\mathcal{L}}_{\mu'} = \bar{\mathcal{L}}_{\mu'} , \tag{2.99}$$

we get the equation for the matrix H

$$(D+2)H^{\mu'}{}_{\nu'} = -\bar{\gamma}^{\mu'}{}_{\nu'} . \tag{2.100}$$

Substituting (2.92) and (2.95) in (2.100), we obtain an ordinary first order differential equation

$$\left(t\frac{d}{dt} + \hat{1}\cdot 2\right)H = \frac{\sin t}{t} , \qquad t \equiv \sqrt{\bar{S}} . \tag{2.101}$$

The solution of this equation has the form

$$H = \bar{S}^{-1}\left(\hat{1} - \cos\sqrt{\bar{S}} + C_3\right) , \tag{2.102}$$

where C_3 is the integration constant. Using the initial condition $[\bar{\mathcal{A}}_{\mu'}] = 0$, we find $C_3 = 0$, i.e., finally

$$H = \bar{S}^{-1}\left(\hat{1} - \cos\sqrt{\bar{S}}\right) ,$$

$$\bar{\mathcal{A}}_{\mu'} = \left\{ \bar{S}^{-1} \left(\hat{1} - \cos\sqrt{\bar{S}} \right) \right\}_{\mu'\nu'} \bar{\mathcal{L}}^{\nu'} . \tag{2.103}$$

Let us apply the formulas (2.95)–(2.97) and (2.103) to the case of the De Sitter space with the curvature

$$R^{\mu}{}_{\alpha\nu\beta} = \Lambda(\delta^{\mu}_{\nu} g_{\alpha\beta} - \delta^{\mu}_{\beta} g_{\alpha\nu}) , \qquad \Lambda = \text{const} . \tag{2.104}$$

The matrix $\bar{S}$, (2.60), (2.61), and the vector $\bar{\mathcal{L}}_{\mu'}$, (2.82), have in this case the form

$$\bar{S}^{\mu'}{}_{\nu'} = 2\Lambda\sigma \Pi_{\perp}{}^{\mu'}{}_{\nu'} , \qquad \bar{\mathcal{L}}_{\mu'} = -\mathcal{R}_{\mu'\nu'}\sigma^{\nu'} , \tag{2.105}$$

where

$$\Pi_{\perp}{}^{\mu'}{}_{\nu'} = \delta^{\mu'}{}_{\nu'} - \frac{\sigma^{\mu'}\sigma_{\nu'}}{2\sigma} , \qquad \Pi_{\perp}{}^{2} = \Pi_{\perp} , \tag{2.106}$$

$$\Pi_{\perp}{}^{\nu'}{}_{\mu'}\sigma^{\mu'} = \sigma_{\nu'}\Pi_{\perp}{}^{\nu'}{}_{\mu'} = 0 ,$$

$\Pi_{\perp}$ being the projector on the hypersurface that is orthogonal to the vector $\sigma^{\mu'}$. Using (2.106) we obtain for an analytic function of the matrix $\bar{S}$

$$f(\bar{S}) = \sum_{k \geq 0} c_k \bar{S}^k = f(0)(\hat{1} - \Pi_{\perp}) + f(2\Lambda\sigma)\Pi_{\perp} . \tag{2.107}$$

Thus the formal expressions (2.95)–(2.97) and (2.103) take the concrete form

$$\begin{aligned}
\bar{\gamma}^{\mu'}{}_{\nu'} &= -\delta^{\mu'}{}_{\nu'}\Phi + \frac{\sigma^{\mu'}\sigma_{\nu'}}{2\sigma}(\Phi - 1) , \\
\bar{\eta}^{\mu'}{}_{\nu'} &= -\delta^{\mu'}{}_{\nu'}\Phi^{-1} + \frac{\sigma^{\mu'}\sigma_{\nu'}}{2\sigma}(\Phi^{-1} - 1) , \\
X^{\mu'\nu'} &= g^{\mu'\nu'}\Phi^{-2} + \frac{\sigma^{\mu'}\sigma^{\nu'}}{2\sigma}(1 - \Phi^{-2}) , \\
H^{\mu'}{}_{\nu'} &= \delta^{\mu'}{}_{\nu'}\Psi + \frac{\sigma^{\mu'}\sigma_{\nu'}}{2\sigma}\left(\frac{1}{2} - \Psi\right) ,
\end{aligned} \tag{2.108}$$

where

$$\Phi = \frac{\sin\sqrt{2\Lambda\sigma}}{\sqrt{2\Lambda\sigma}} , \qquad \Psi = \frac{1 - \cos\sqrt{2\Lambda\sigma}}{2\Lambda\sigma} .$$

Using (2.105) one can neglect the longitudinal terms in the matrix H (2.108) and obtain the quantity $\bar{\mathcal{A}}_{\mu'}$:

$$\bar{\mathcal{A}}_{\mu'} = -\Psi \mathcal{R}_{\mu'\nu'}\sigma^{\nu'} . \tag{2.109}$$

In the case (2.90) it is transverse,

$$\bar{\nabla}^{\mu'}\bar{\mathcal{A}}_{\mu'} = 0 . \tag{2.110}$$

Let us stress that all formulas obtained in present section are valid for spaces of any dimension and signature.

2.3 Technique for Calculation of De Witt Coefficients

Let us apply the method of covariant expansions developed above to the calculation of the De Witt coefficients a_k, i.e., the coefficients of the asymptotic expansion of the transfer function (1.43). Below we follow our papers [6, 12, 11, 9].

From the recurrence relations (1.45) we obtain the formal solution

$$a_k = \left(1 + \frac{1}{k}D\right)^{-1} F \left(1 + \frac{1}{k-1}D\right)^{-1} F \cdots (1+D)^{-1} F\,, \tag{2.111}$$

where the operator D is defined by (2.4) and the operator F has the form

$$F = \mathcal{P}^{-1}(\hat{1}\,\Delta^{-1/2}\,\Box\,\Delta^{1/2} + Q)\mathcal{P}\,. \tag{2.112}$$

In the present section we develop a convenient covariant and effective method that gives a practical meaning to the formal expression (2.111). It will suffice, in particular, to calculate the coincidence limits of the De Witt coefficients a_k and their derivatives.

First of all, we suppose that there exist finite coincidence limits of the De Witt coefficients

$$[a_k] \equiv \lim_{x \to x'} a_k(x, x')\,, \tag{2.113}$$

that do not depend on the way how the points x and x' approach each other, i.e., the De Witt coefficients $a_k(x, x')$ are analytical functions of the coordinates of the point x near the point x'. Otherwise, i.e., if the limit (2.113) is singular, the operator $(1+\frac{1}{k}D)$ does not have a single-valued inverse operator, since there exist the eigenfunctions of this operator with eigenvalues equal to zero

$$D\sigma^{-n/2}|n>= 0, \qquad \left(1 + \frac{1}{k}D\right)\sigma^{-(k+n)/2}|n>= 0\,. \tag{2.114}$$

Therefore, the zeroth coefficient $a_0(x, x')$ is defined in general up to an arbitrary function $f(\sigma^{\mu'}/\sqrt{\sigma}; x')$, and the inverse operator $(1+\frac{1}{k}D)^{-1}$ is defined up to an arbitrary function $\sigma^{-k/2} f_k(\sigma^{\mu'}/\sqrt{\sigma}; x')$.

Using the covariant Taylor series (2.21) and (2.32) for the De Witt coefficients

$$a_k = \sum_{n \geq 0} |n>< n|a_k> \tag{2.115}$$

and defining the inverse operator $(1+\frac{1}{k}D)^{-1}$ in form of the eigenfunctions expansion (2.35),

$$\left(1 + \frac{1}{k}D\right)^{-1} = \sum_{n \geq 0} \frac{k}{k+n} |n>< n|\,, \tag{2.116}$$

we obtain from (2.111)

$$< n|a_k > = \sum_{n_1,\cdots,n_{k-1}\geq 0} \frac{k}{k+n} \cdot \frac{k-1}{k-1+n_{k-1}} \cdots \frac{1}{1+n_1}$$

$$\times < n|F|n_{k-1} >< n_{k-1}|F|n_{k-2} > \cdots < n_1|F|0 > , \quad (2.117)$$

where

$$< m|F|n >= \left[\nabla_{(\mu_1} \cdots \nabla_{\mu_m)} F \frac{(-1)^n}{n!} \sigma^{\nu'_1} \cdots \sigma^{\nu'_n}\right] \quad (2.118)$$

are the matrix elements (2.36) of the operator F, (2.112).

Since the operator F, (2.112), is a differential operator of second order, its matrix elements $< m|F|n >$, (2.118), do not vanish only for $n \leq m+2$. Therefore, the sum (2.117) always contains a finite number of terms, i.e., the summation over n_i is limited from above

$$n_1 \geq 0 , \qquad n_i \leq n_{i+1} + 2 , \qquad (i = 1, \ldots, k-1;\ n_k \equiv n) . \quad (2.119)$$

Thus we reduced the problem of calculation of the De Witt coefficients to the calculation of the matrix elements (2.118) of the operator F, (2.112). For the calculation of the matrix elements (2.118) it is convenient to write the operator F, (2.112), in terms of the operators $\bar\nabla_{\mu'}$, (2.41). Using (2.112), (2.41)–(2.46) and (2.51) we obtain

$$\begin{aligned} F &= \mathcal{P}^{-1} \Delta^{1/2} \bar\nabla_{\mu'} \Delta^{-1} X^{\mu'\nu'} \bar\nabla_{\nu'} \Delta^{1/2} \mathcal{P} + \bar Q \\ &= \left\{\hat 1 (\bar\nabla_{\mu'} - \zeta_{\mu'}) + \bar{\mathcal{A}}_{\mu'}\right\} X^{\mu'\nu'} \left\{\hat 1 (\bar\nabla_{\nu'} + \zeta_{\nu'}) + \bar{\mathcal{A}}_{\nu'}\right\} + \bar Q \\ &= \hat 1 X^{\mu'\nu'} \bar\nabla_{\mu'} \bar\nabla_{\nu'} + Y^{\mu'} \bar\nabla_{\mu'} + Z , \end{aligned} \quad (2.120)$$

where

$$\bar Q = \mathcal{P}^{-1} Q \mathcal{P} ,$$

$$\zeta_{\mu'} = \bar\nabla_{\mu'} \zeta = \bar\nabla_{\mu'} \log \Delta^{1/2} , \quad (2.121)$$

$$Y^{\mu'} = \hat 1 \bar\nabla_{\nu'} X^{\nu'\mu'} + 2 X^{\mu'\nu'} \bar{\mathcal{A}}_{\nu'} , \quad (2.122)$$

$$Z = X^{\mu'\nu'} \left(\bar{\mathcal{A}}_{\mu'} \bar{\mathcal{A}}_{\nu'} - \hat 1 \zeta_{\mu'} \zeta_{\nu'}\right) + \bar\nabla_{\nu'} \left\{X^{\mu'\nu'} \left(\hat 1 \zeta_{\mu'} + \bar{\mathcal{A}}_{\mu'}\right)\right\} + \bar Q . \quad (2.123)$$

Now one can easily calculate the matrix elements (2.118). Using (2.54), (2.5) and (2.48) we obtain from (2.118)

$$< m|F|m+2 >= \hat 1 \delta^{(\nu_1 \cdots \nu_m}_{\mu_1 \cdots \mu_m} g^{\nu_{m+1} \nu_{m+2})} ,$$

$$< m|F|m+1 >= 0 ,$$

$$
< m|F|n > = \binom{m}{n} \delta^{\nu_1\cdots\nu_n}_{(\mu_1\cdots\mu_n} Z_{\mu_{n+1}\cdots\mu_m)} - \binom{m}{n-1} \delta^{(\nu_1\cdots\nu_{n-1}}_{(\mu_1\cdots\mu_{n-1}} Y^{\nu_n)}{}_{\mu_n\cdots\mu_m)}
$$

$$
+ \binom{m}{n-2} \hat{1}\, \delta^{(\nu_1\cdots\nu_{n-2}}_{(\mu_1\cdots\mu_{n-2}} X^{\nu_{n-1}\nu_n)}{}_{\mu_{n-1}\cdots\mu_m)} , \qquad (2.124)
$$

where

$$
Z_{\mu_1\cdots\mu_n} = (-1)^n \left[\bar{\nabla}_{(\mu'_1} \cdots \bar{\nabla}_{\mu'_n)} Z \right] ,
$$

$$
Y^{\nu}{}_{\mu_1\cdots\mu_n} = (-1)^n \left[\bar{\nabla}_{(\mu'_1} \cdots \bar{\nabla}_{\mu'_n)} Y^{\nu'} \right] ,
$$

$$
X^{\nu_1\nu_2}{}_{\mu_1\cdots\mu_n} = (-1)^n \left[\bar{\nabla}_{(\mu'_1} \cdots \bar{\nabla}_{\mu'_n)} X^{\nu'_1\nu'_2} \right] . \qquad (2.125)
$$

These In (2.124) it is meant that the binomial coefficient $\binom{n}{k}$ is equal to zero if $k < 0$ or $n < k$.

Thus, to calculate the matrix elements (2.118) it is sufficient to have the coincidence limits of the symmetrized covariant derivatives (2.125) of the coefficient functions $X^{\mu'\nu'}$, $Y^{\mu'}$ and Z, (2.122), (2.123), i.e., the coefficients of their Taylor expansions (2.48), that are expressed in terms of the Taylor coefficients of the quantities $X^{\mu'\nu'}$, (2.76), $\bar{A}_{\mu'}$, (2.88), and ζ, (2.78), (2.79), found in the Sect. 2.2. These expressions are computed explicitly in [9, 11, 12]. From the dimensional arguments it is obvious that for $m = n$ the matrix elements $< m|F|n >$, (2.124), (2.125), are expressed in terms of the curvature tensors $R^{\alpha}{}_{\beta\gamma\delta}$, $\mathcal{R}_{\mu\nu}$ and the matrix Q; for $m = n + 1$ – in terms of the quantities ∇R, $\nabla\mathcal{R}$ and ∇Q; for $m = n + 2$ – in terms of the quantities of the form R^2, $\nabla\nabla R$ etc.

In the calculation of the De Witt coefficients by means of the matrix algorithm (2.117) a "diagrammatic" technique, i.e., a graphic method for enumerating the different terms of the sum (2.117), turns out to be very convenient and pictorial. The matrix elements $< m|F|n >$, (2.118), are presented by some blocks with m lines coming in from the left and n lines going out to the right (Fig. 1),

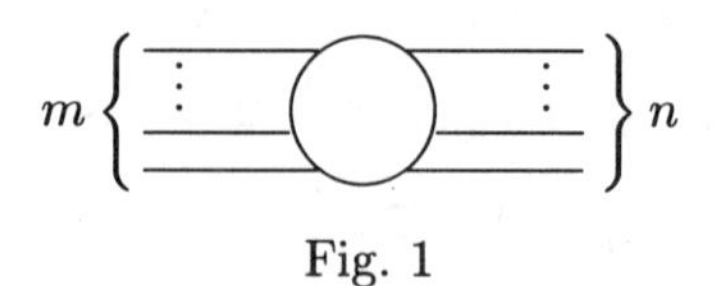

Fig. 1

and the product of the matrix elements $< m|F|k >< k|F|n >$ — by two blocks connected by k intermediate lines (Fig. 2),

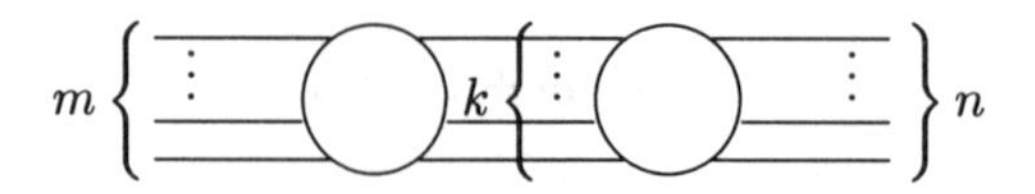

Fig. 2

that represents the contractions of corresponding tensor indices.

To obtain the coefficient $< n|a_k >$, (2.117), one should draw all possible diagrams with k blocks connected in all possible ways by any number of intermediate lines. When doing this, one should keep in mind that the number of the lines, going out of any block, cannot be greater than the number of the lines, coming in, by more than two and by exactly one (see (2.124)). Then one should sum up all diagrams with the weight determined for each diagram by the number of intermediate lines from (2.117). Drawing of such diagrams is of no difficulties. Therefore, the main problem is reduced to the calculation of some standard blocks.

Note that the elaborated technique does not depend at all on the dimension of the space-time and on the signature of the metric and enables one to obtain results in most general case. It is also very algorithmic and can be easily adopted for symbolic computer calculations [44].

2.4 De Witt Coefficients a_3 and a_4

Using the developed technique one can calculate the coincidence limits (2.113) of the De Witt coefficients $[a_3]$ and $[a_4]$. The coefficients $[a_1]$ and $[a_2]$, that determine the one-loop divergences (1.51), (1.53), have been calculated long ago by De Witt [80], Christensen [62, 63] and Gilkey [132]. The coefficient $[a_3]$, that describes the vacuum polarization of massive quantum fields in the lowest non-vanishing approximation $\sim 1/m^2$, (1.54), was calculated in general form first by Gilkey [132]. The coefficient $[a_4]$ in general form has been calculated in the papers [12, 11, 9, 4]. The papers [24, 21]) provide comprehensive reviews on the calculation of the coefficients a_k as well as and more complete bibliography and historical comments as to what was computed where for the first time.

The diagrams for the De Witt coefficients $[a_3]$ and $[a_4]$ have the form, (2.117),

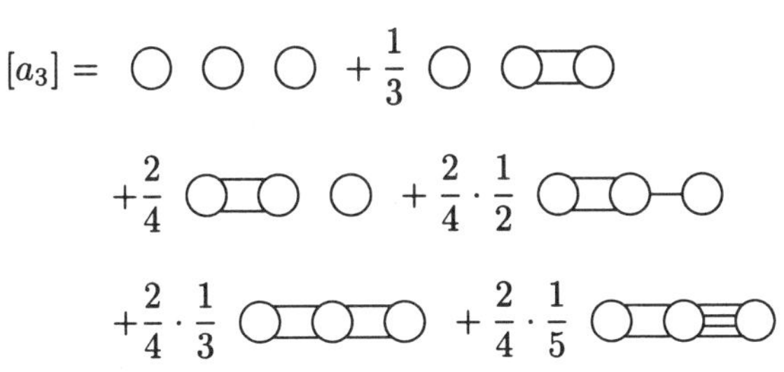

$$
\begin{aligned}
[a_3] = {} & \bigcirc\ \bigcirc\ \bigcirc + \frac{1}{3}\ \bigcirc\ \bigcirc\!=\!\bigcirc \\
& + \frac{2}{4}\ \bigcirc\!=\!\bigcirc\ \bigcirc + \frac{2}{4}\cdot\frac{1}{2}\ \bigcirc\!=\!\bigcirc\!-\!\bigcirc \\
& + \frac{2}{4}\cdot\frac{1}{3}\ \bigcirc\!=\!\bigcirc\!=\!\bigcirc + \frac{2}{4}\cdot\frac{1}{5}\ \bigcirc\!=\!\bigcirc\!\equiv\!\bigcirc\ ,
\end{aligned}
\qquad (2.126)
$$

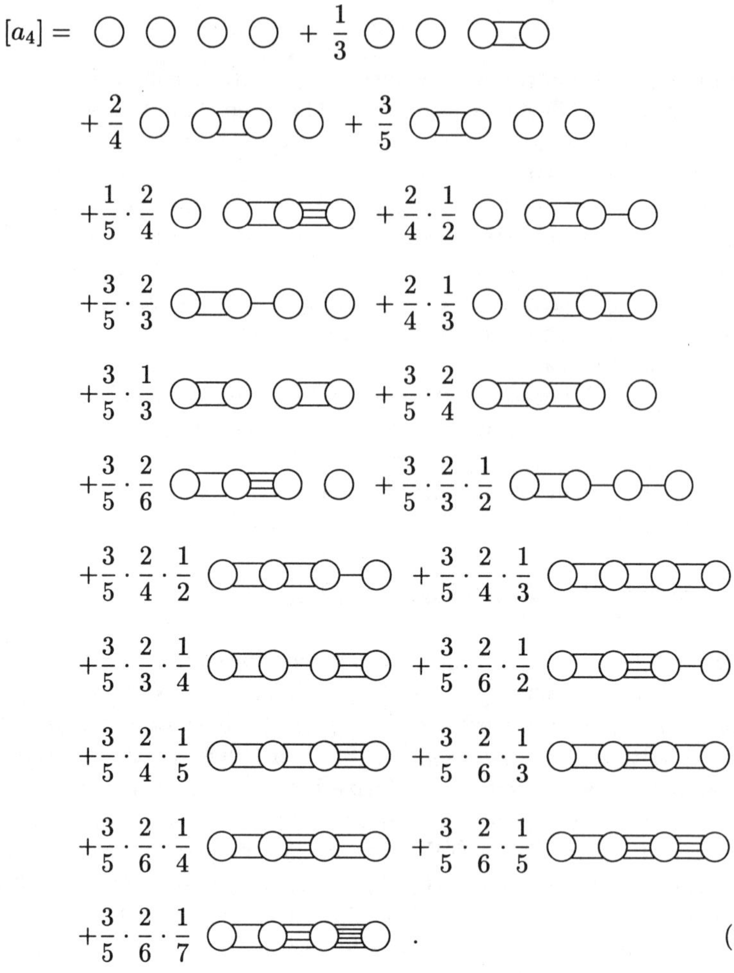

$$[a_4] = \quad + \frac{1}{3} \quad + \frac{2}{4} \quad + \frac{3}{5}$$

$$+ \frac{1}{5}\cdot\frac{2}{4} \quad + \frac{2}{4}\cdot\frac{1}{2}$$

$$+ \frac{3}{5}\cdot\frac{2}{3} \quad + \frac{2}{4}\cdot\frac{1}{3}$$

$$+ \frac{3}{5}\cdot\frac{1}{3} \quad + \frac{3}{5}\cdot\frac{2}{4}$$

$$+ \frac{3}{5}\cdot\frac{2}{6} \quad + \frac{3}{5}\cdot\frac{2}{3}\cdot\frac{1}{2}$$

$$+ \frac{3}{5}\cdot\frac{2}{4}\cdot\frac{1}{2} \quad + \frac{3}{5}\cdot\frac{2}{4}\cdot\frac{1}{3}$$

$$+ \frac{3}{5}\cdot\frac{2}{3}\cdot\frac{1}{4} \quad + \frac{3}{5}\cdot\frac{2}{6}\cdot\frac{1}{2}$$

$$+ \frac{3}{5}\cdot\frac{2}{4}\cdot\frac{1}{5} \quad + \frac{3}{5}\cdot\frac{2}{6}\cdot\frac{1}{3}$$

$$+ \frac{3}{5}\cdot\frac{2}{6}\cdot\frac{1}{4} \quad + \frac{3}{5}\cdot\frac{2}{6}\cdot\frac{1}{5}$$

$$+ \frac{3}{5}\cdot\frac{2}{6}\cdot\frac{1}{7} \quad . \tag{2.127}$$

Substituting the matrix elements (2.124) in (2.126) and (2.127) we express the coefficients $[a_3]$ and $[a_4]$ in terms of the quantities X, Y and Z, (2.125),

$$[a_3] = P^3 + \frac{1}{2}[P, Z_{(2)}]_+ + \frac{1}{2}B^\mu Z_\mu + \frac{1}{10}Z_{(4)} + \delta_3 \ , \tag{2.128}$$

$$\begin{aligned}[a_4] = {} & P^4 + \frac{3}{5}\left[P^2, Z_{(2)}\right]_+ + \frac{4}{5}PZ_{(2)}P + \frac{4}{5}[P, B^\mu Z_\mu]_+ \\ & + \frac{2}{5}B^\mu P Z_\mu - \frac{2}{5}B_\mu Y^{\nu\mu} Z_\nu + \frac{1}{3}Z_{(2)}Z_{(2)} + \frac{2}{5}B^\mu Z_{\mu(2)} \\ & + \frac{2}{5}C^\mu Z_\mu + \frac{1}{5}\left[P, Z_{(4)}\right]_+ + \frac{4}{15}\mathcal{D}^{\mu\nu} Z_{\mu\nu} + \frac{1}{35}Z_{(6)} + \delta_4 \ , \end{aligned} \tag{2.129}$$

where

$$\delta_3 = \frac{1}{6} U_1^{\mu\nu} Z_{\mu\nu} , \tag{2.130}$$

$$\delta_4 = \frac{4}{15} P U_1^{\mu\nu} Z_{\mu\nu} + U_1^{\mu\nu} \left\{ \frac{3}{10} Z_{\mu\nu} P + \frac{1}{10} P Z_{\mu\nu} + \frac{3}{10} Z_\mu Z_\nu \right.$$

$$\left. - \frac{3}{10} Y^\alpha{}_{\mu\nu} Z_\alpha - \frac{1}{5} Y^\alpha{}_\mu Z_{\alpha\nu} + \frac{1}{10} X^{\alpha\beta}{}_{\mu\nu} Z_{\alpha\beta} + \frac{7}{50} Z_{\mu\nu(2)} \right\}$$

$$+ \frac{1}{5} U_2^{\mu\nu\lambda} Z_{\mu\nu\lambda} + \frac{4}{25} U_3^{\mu\nu\alpha\beta} Z_{\mu\nu\alpha\beta} . \tag{2.131}$$

Here the following notation is introduced:

$$P = [Z] ,$$

$$Z_{(2)} = g^{\mu_1\mu_2} Z_{\mu_1\mu_2} , \tag{2.132}$$

$$Z_{(4)} = g^{\mu_1\mu_2} g^{\mu_3\mu_4} Z_{\mu_1\cdots\mu_4} ,$$

$$Z_{(6)} = g^{\mu_1\mu_2} g^{\mu_3\mu_4} g^{\mu_5\mu_6} Z_{\mu_1\cdots\mu_6} , \tag{2.133}$$

$$Z_{\mu(2)} = g^{\mu_1\mu_2} Z_{\mu\mu_1\mu_2} ,$$

$$Z_{\mu\nu(2)} = g^{\mu_1\mu_2} Z_{\mu\nu\mu_1\mu_2} , \tag{2.134}$$

$$B^\mu = Z^\mu - \frac{1}{2} g^{\mu_1\mu_2} Y^\mu{}_{\mu_1\mu_2} , \tag{2.135}$$

$$C^\mu = Z^\mu{}_{(2)} - \frac{1}{4} g^{\mu_1\mu_2} g^{\mu_3\mu_4} Y^\mu{}_{\mu_1\cdots\mu_4} , \tag{2.136}$$

$$\mathcal{D}^{\mu\nu} = Z^{\mu\nu} - g^{\mu_1\mu_2} Y^{(\mu\nu)}{}_{\mu_1\mu_2} + \frac{1}{4} \hat{1}\, g^{\mu_1\mu_2} g^{\mu_3\mu_4} X^{\mu\nu}{}_{\mu_1\cdots\mu_4} , \tag{2.137}$$

$$U_1^{\mu\nu} = -2 Y^{(\mu\nu)} + \hat{1}\, g^{\mu_1\mu_2} X^{\mu\nu}{}_{\mu_1\mu_2} , \tag{2.138}$$

$$U_2^{\mu\nu\lambda} = -Y^{(\mu\nu\lambda)} + \hat{1}\, g^{\mu_1\mu_2} X^{(\mu\nu\lambda)}{}_{\mu_1\mu_2} , \tag{2.139}$$

$$U_3^{\mu\nu\alpha\beta} = X^{(\mu\nu\alpha\beta)} , \tag{2.140}$$

and $[A, B]_+$ denotes the anti-commutator of the matrices A and B.

Using the formulas (2.132)–(2.140), (2.125), (2.122) and (2.123) and the quantities $X^{\mu'\nu'}$, $\bar{A}_{\mu'}$ and ζ, (2.70)–(2.89), and omitting cumbersome computations we obtain

$$P = Q + \frac{1}{6} \hat{1}\, R , \tag{2.141}$$

$$Y_{\mu\nu} = \mathcal{R}_{\mu\nu} + \frac{1}{3} \hat{1}\, R_{\mu\nu} , \tag{2.142}$$

$$U_1^{\mu\nu} = U_3^{\mu\nu\alpha\beta} = 0 \,, \tag{2.143}$$

$$Z_\mu = \nabla_\mu P - \frac{1}{3} J_\mu \,, \tag{2.144}$$

$$B_\mu = \nabla_\mu P + \frac{1}{3} J_\mu \,, \tag{2.145}$$

$$U_2^{\mu\nu\lambda} = 0 \,, \tag{2.146}$$

$$Z_{(2)} = \Box \left(Q + \frac{1}{5} \hat{1}\, R \right) + \frac{1}{30} \hat{1}\, \left(R_{\mu\nu\alpha\beta} R^{\mu\nu\alpha\beta} - R_{\mu\nu} R^{\mu\nu} \right) + \frac{1}{2} \mathcal{R}_{\mu\nu} \mathcal{R}^{\mu\nu} \,, \tag{2.147}$$

$$Z_{\mu\nu} = W_{\mu\nu} - \frac{1}{2} \nabla_{(\mu} J_{\nu)} \,, \tag{2.148}$$

$$\mathcal{D}_{\mu\nu} = W_{\mu\nu} + \frac{1}{2} \nabla_{(\mu} J_{\nu)} \,, \tag{2.149}$$

$$Z_{\mu(2)} = V_\mu + G_\mu \,, \tag{2.150}$$

$$C_\mu = V_\mu - G_\mu \,, \tag{2.151}$$

where

$$J_\mu = \nabla_\alpha \mathcal{R}^\alpha{}_\mu \,, \tag{2.152}$$

$$W_{\mu\nu} = \nabla_{(\mu} \nabla_{\nu)} \left(Q + \frac{3}{20} \hat{1}\, R \right) + \hat{1} \left\{ \frac{1}{20} \Box R_{\mu\nu} - \frac{1}{15} R_{\mu\alpha} R^\alpha{}_\nu \right.$$

$$\left. + \frac{1}{30} R_{\mu\alpha\beta\gamma} R_\nu{}^{\alpha\beta\gamma} + \frac{1}{30} R_{\alpha\beta} R^\alpha{}_\mu{}^\beta{}_\nu \right\} + \frac{1}{2} \mathcal{R}_{\alpha(\mu} \mathcal{R}^\alpha{}_{\nu)} \,, \tag{2.153}$$

$$V_\mu = Q_{\mu(2)} + \frac{1}{2} [\mathcal{R}_{\alpha\beta}, \nabla_\mu \mathcal{R}^{\alpha\beta}]_+ - \frac{1}{3} [J^\nu, \mathcal{R}_{\nu\mu}]_+$$

$$+ \hat{1} \left\{ \frac{1}{5} \nabla_\mu \Box R + \frac{1}{9} R^\nu_\mu \nabla_\nu R + \frac{1}{15} R_{\alpha\beta\gamma\delta} \nabla_\mu R^{\alpha\beta\gamma\delta} \right.$$

$$\left. - \frac{1}{15} R_{\alpha\beta} \nabla_\mu R^{\alpha\beta} \right\} \,, \tag{2.154}$$

$$Q_{\mu(2)} = g^{\mu_1\mu_2} \nabla_{(\mu} \nabla_{\mu_1} \nabla_{\mu_2)} Q$$

$$= \nabla_\mu \Box Q + [\mathcal{R}_{\nu\mu}, \nabla^\nu Q] + \frac{1}{3} [J_\mu, Q] + \frac{2}{3} R^\nu_\mu \nabla_\nu Q \,, \tag{2.155}$$

$$G_\mu = -\frac{1}{5}\Box J_\mu - \frac{2}{15}[\mathcal{R}_{\alpha\mu}, J^\alpha] - \frac{1}{10}[\mathcal{R}_{\alpha\beta}, \nabla_\mu \mathcal{R}^{\alpha\beta}]$$

$$-\frac{2}{15}R^{\alpha\beta}\nabla_\alpha \mathcal{R}_{\beta\mu} + \frac{2}{15}R_{\mu\alpha\beta\gamma}\nabla^\alpha \mathcal{R}^{\beta\gamma} - \frac{7}{45}R_{\mu\alpha}J^\alpha$$

$$+\frac{2}{5}\nabla_\alpha R_{\beta\mu}\mathcal{R}^{\beta\alpha} - \frac{1}{15}\nabla^\alpha R \mathcal{R}_{\alpha\mu} . \tag{2.156}$$

The result for the coefficient $Z_{(4)}$ has more complicated form

$$Z_{(4)} = Q_{(4)} + 2[\mathcal{R}^{\mu\nu}, \nabla_\mu J_\nu]_+ + \frac{8}{9}J_\mu J^\mu + \frac{4}{3}\nabla_\mu \mathcal{R}_{\alpha\beta}\nabla^\mu \mathcal{R}^{\alpha\beta}$$

$$+6\mathcal{R}_{\mu\nu}\mathcal{R}^\nu{}_\gamma \mathcal{R}^{\gamma\mu} + \frac{10}{3}R^{\alpha\beta}\mathcal{R}^\mu{}_\alpha \mathcal{R}_{\mu\beta} - R^{\mu\nu\alpha\beta}\mathcal{R}_{\mu\nu}\mathcal{R}_{\alpha\beta}$$

$$+\hat{1}\left\{\frac{3}{14}\Box^2 R + \frac{1}{7}R^{\mu\nu}\nabla_\mu\nabla_\nu R - \frac{2}{21}R^{\mu\nu}\Box R_{\mu\nu}\right.$$

$$+\frac{4}{7}R^\alpha{}_\mu{}^\beta{}_\nu \nabla_\alpha\nabla_\beta R^{\mu\nu} + \frac{4}{63}\nabla_\mu R\nabla^\mu R - \frac{1}{42}\nabla_\mu R_{\alpha\beta}\nabla^\mu R^{\alpha\beta}$$

$$-\frac{1}{21}\nabla_\mu R_{\alpha\beta}\nabla^\alpha R^{\beta\mu} + \frac{3}{28}\nabla_\mu R_{\alpha\beta\gamma\delta}\nabla^\mu R^{\alpha\beta\gamma\delta} + \frac{2}{189}R^\alpha_\beta R^\beta_\gamma R^\gamma_\alpha$$

$$-\frac{2}{63}R_{\alpha\beta}R^{\mu\nu}R^\alpha{}_\mu{}^\beta{}_\nu + \frac{2}{9}R_{\alpha\beta}R^\alpha{}_{\mu\nu\lambda}R^{\beta\mu\nu\lambda}$$

$$\left.-\frac{16}{189}R_{\alpha\beta}{}^{\mu\nu}R_{\mu\nu}{}^{\sigma\rho}R_{\sigma\rho}{}^{\alpha\beta} - \frac{88}{189}R^\alpha{}_\mu{}^\beta{}_\nu R^\mu{}_\sigma{}^\nu{}_\rho R^\sigma{}_\alpha{}^\rho{}_\beta\right\} , \tag{2.157}$$

where

$$Q_{(4)} = g^{\mu_1\mu_2}g^{\mu_3\mu_4}\nabla_{(\mu_1}\cdots\nabla_{\mu_4)}Q$$

$$= \Box^2 Q - \frac{1}{2}[\mathcal{R}^{\mu\nu}, [\mathcal{R}_{\mu\nu}, Q]] - \frac{2}{3}[J^\mu, \nabla_\mu Q]$$

$$+\frac{2}{3}R^{\mu\nu}\nabla_\mu\nabla_\nu Q + \frac{1}{3}\nabla_\mu R\nabla^\mu Q . \tag{2.158}$$

From the fact that the quantities $U_1^{\mu\nu}$, $U_2^{\mu\nu\lambda}$ and $U_3^{\mu\nu\lambda\sigma}$, (2.138)–(2.140), are equal to zero, (2.143), (2.146), it follows that the quantities δ_3 and δ_4, (2.130), (2.131), are equal to zero too:

$$\delta_3 = \delta_4 = 0 . \tag{2.159}$$

To calculate the De Witt coefficient $[a_3]$, (2.128), it is sufficient to have the formulas listed above. This coefficient is presented explicitly in the paper [132]. (Let us note, to avoid misunderstanding, that our De Witt coefficients a_k, (1.43), differ from the coefficients $\tilde{a}_k$ used by the other authors [80, 62, 63, 132] by a factor: $a_k = k!\tilde{a}_k$.)

Let us calculate the De Witt coefficient A_3, (1.55), that determines the renormalized one-loop effective action (1.54) in the four-dimensional physical space-time in the lowest non-vanishing approximation $\sim 1/m^2$ [120, 121]. By integrating by parts and omitting the total derivatives we obtain from (2.128), (2.141), (2.144), (2.145), (2.147) and (2.157)–(2.159)

$$
\begin{aligned}
A_3 = \int d^n x\, g^{1/2} \mathrm{str}\Bigg\{ & P^3 + \frac{1}{30} P\Big(R_{\mu\nu\alpha\beta}R^{\mu\nu\alpha\beta} - R_{\mu\nu}R^{\mu\nu} + \Box R\Big) \\
& + \frac{1}{2} P \mathcal{R}_{\mu\nu}\mathcal{R}^{\mu\nu} + \frac{1}{2} P \Box P - \frac{1}{10} J_\mu J^\mu \\
& + \frac{1}{30}\left(2\mathcal{R}^\mu{}_\nu \mathcal{R}^\nu{}_\alpha \mathcal{R}^\alpha{}_\mu - 2R^\mu_\nu \mathcal{R}_{\mu\alpha}\mathcal{R}^{\alpha\nu} + R^{\mu\nu\alpha\beta}\mathcal{R}_{\mu\nu}\mathcal{R}_{\alpha\beta}\right) \\
& + \hat{1}\Bigg[-\frac{1}{630} R \Box R + \frac{1}{140} R_{\mu\nu} \Box R^{\mu\nu} + \frac{1}{7560}\Big(-64 R^\mu_\nu R^\nu_\lambda R^\lambda_\mu \\
& + 48 R^{\mu\nu} R_{\alpha\beta} R^\alpha{}_\mu{}^\beta{}_\nu + 6 R_{\mu\nu} R^\mu{}_{\alpha\beta\gamma} R^{\nu\alpha\beta\gamma} \\
& + 17 R_{\mu\nu}{}^{\alpha\beta} R_{\alpha\beta}{}^{\sigma\rho} R_{\sigma\rho}{}^{\mu\nu} - 28 R^\mu{}_\alpha{}^\nu{}_\beta R^\alpha{}_\sigma{}^\beta{}_\rho R^\sigma{}_\mu{}^\rho{}_\nu \Big)\Bigg]\Bigg\} . \qquad (2.160)
\end{aligned}
$$

The formula (2.160) is valid for any dimension of the space and for any fields.

To obtain the explicit expression for the coefficient $[a_4]$ one has to substitute (2.141)–(2.159) in (2.129) as well as to calculate the quantity $Z_{(6)}$, (2.133). To write down this quantity in a compact way we define the following tensors constructed from the covariant derivatives of the curvature tensor:

$$
\begin{aligned}
I^{\alpha\beta}{}_{\gamma\mu_1\cdots\mu_n} &= \nabla_{(\mu_1} \cdots \nabla_{\mu_{n-1}} R^\alpha{}_{|\gamma|}{}^\beta{}_{\mu_n)} , \\
K^{\alpha\beta}{}_{\mu_1\cdots\mu_n} &= \nabla_{(\mu_1} \cdots \nabla_{\mu_{n-2}} R^\alpha{}_{\mu_{n-1}}{}^\beta{}_{\mu_n)} , \\
L^\alpha{}_{\mu_1\cdots\mu_n} &= \nabla_{(\mu_1} \cdots \nabla_{\mu_{n-1}} R^\alpha{}_{\mu_n)} , \qquad (2.161) \\
M_{\mu_1\cdots\mu_n} &= \nabla_{(\mu_1} \cdots \nabla_{\mu_{n-2}} R_{\mu_{n-1}\mu_n)} , \\
\mathcal{R}^\mu{}_{\mu_1\cdots\mu_n} &= \nabla_{(\mu_1} \cdots \nabla_{\mu_{n-1}} \mathcal{R}^\mu{}_{\mu_n)} ,
\end{aligned}
$$

and denote the contracted symmetrized covariant derivatives just by a number in the brackets (analogously to (2.132)–(2.134). For example,

$$
\begin{aligned}
I^{\alpha\beta}{}_{\gamma\mu(2)} &= g^{\mu_1\mu_2} I^{\alpha\beta}{}_{\gamma\mu\mu_1\mu_2} \,, \\
K^{\alpha\beta}{}_{\mu\nu(4)} &= g^{\mu_1\mu_2} g^{\mu_3\mu_4} K^{\alpha\beta}{}_{\mu\nu\mu_1\cdots\mu_4} \,, \\
L^{\alpha}{}_{\mu(4)} &= g^{\mu_1\mu_2} g^{\mu_3\mu_4} L^{\alpha}{}_{\mu\mu_1\cdots\mu_4} \,, \\
M_{(8)} &= g^{\mu_1\mu_2} \cdots g^{\mu_7\mu_8} M_{\mu_1\cdots\mu_8} \,, \\
\mathcal{R}^{\mu}{}_{\nu(4)} &= g^{\mu_1\mu_2} g^{\mu_3\mu_4} \mathcal{R}^{\mu}{}_{\nu\mu_1\cdots\mu_4} \,,
\end{aligned}
\tag{2.162}
$$

etc.

In terms of introduced quantities, (2.161), and the notation (2.162) the quantity $Z_{(6)}$, (2.133), takes the form

$$
Z_{(6)} = Z^M_{(6)} + \hat{1}\, Z^S_{(6)} \,. \tag{2.163}
$$

Here $Z^M_{(6)}$ is the matrix contribution

$$
\begin{aligned}
Z^M_{(6)} = {} & Q_{(6)} + \frac{5}{2}[\mathcal{R}^{\mu\nu}, \mathcal{R}_{\mu\nu(4)}]_+ + \frac{32}{5}[\mathcal{R}^{\mu\nu\alpha}, \mathcal{R}_{\mu\nu\alpha(2)}]_+ \\
& - \frac{8}{5}[J^{\mu}, \mathcal{R}_{\mu(4)}]_+ + \frac{9}{2}\mathcal{R}^{\mu\nu\alpha\beta}\mathcal{R}_{\mu\nu\alpha\beta} + \frac{27}{4}\mathcal{R}^{\mu\nu}{}_{(2)}\mathcal{R}_{\mu\nu(2)} \\
& + \frac{5}{4}R^{\mu}_{\nu}[\mathcal{R}^{\nu\alpha}, \mathcal{R}_{\mu\alpha(2)}]_+ + \frac{5}{2}R_{\mu}{}^{\nu\alpha\beta}[\mathcal{R}^{\mu\gamma}, \mathcal{R}_{\alpha\beta\nu\gamma}]_+ \\
& + \frac{15}{8}R^{\mu\nu\alpha\beta}[\mathcal{R}_{\mu\nu}, \mathcal{R}_{\alpha\beta(2)}]_+ + \frac{44}{15}\nabla_{\mu}R_{\alpha\nu}[\mathcal{R}^{\mu\alpha}, J^{\nu}]_+ \\
& + \frac{22}{5}K^{\mu\nu\alpha\beta\gamma}[\mathcal{R}_{\mu\gamma}, \mathcal{R}_{\nu\alpha\beta}]_+ + \frac{22}{5}K_{\mu\nu\alpha(2)}[\mathcal{R}^{\mu\gamma}, \mathcal{R}^{\nu\alpha}{}_{\gamma}]_+ \\
& + \frac{64}{45}R^{\mu}{}_{\nu}\mathcal{R}_{\mu\alpha\beta}\mathcal{R}^{\nu\alpha\beta} - \frac{16}{15}R^{\mu\nu\alpha\beta}[\nabla_{\beta}\mathcal{R}_{\mu\nu}, J_{\alpha}]_+ \\
& + \frac{256}{45}R_{\mu(\alpha|\nu|\beta)}\mathcal{R}^{\mu\alpha}{}_{\gamma}\mathcal{R}^{\nu\beta\gamma} + \frac{32}{45}R^{\mu\nu}J_{\mu}J_{\nu} \\
& + \left(\frac{6}{5}K_{\mu\nu(4)} + \frac{17}{40}R_{\mu\alpha\beta\gamma}R_{\nu}{}^{\alpha\beta\gamma} + \frac{17}{60}R_{\mu\alpha}R^{\alpha}_{\nu}\right)\mathcal{R}^{\mu\sigma}\mathcal{R}^{\nu}{}_{\sigma} \\
& + \left(\frac{24}{5}K_{\mu\nu\alpha\beta(2)} + \frac{17}{40}R_{\mu\gamma}R^{\gamma}{}_{\beta\nu\alpha} + \frac{17}{40}R_{\nu\gamma}R^{\gamma}{}_{\alpha\mu\beta}\right. \\
& + \frac{17}{30}R_{\mu\nu\sigma\rho}R_{\beta\alpha}{}^{\sigma\rho} + \frac{17}{60}R_{\mu\sigma\alpha\rho}R^{\sigma}{}_{\nu}{}^{\rho}{}_{\beta} \\
& \left. + \frac{51}{80}R_{\mu\beta\sigma\rho}R_{\nu\alpha}{}^{\sigma\rho}\right)\mathcal{R}^{\mu\beta}\mathcal{R}^{\nu\alpha} \,,
\end{aligned}
\tag{2.164}
$$

where

$$Q_{(6)} = g^{\mu_1\mu_2} g^{\mu_3\mu_4} g^{\mu_5\mu_6} \nabla_{(\mu_1} \cdots \nabla_{\mu_6)} Q ,$$

and $Z^S_{(6)}$ is the scalar contribution

$$\begin{aligned}
Z^S_{(6)} &= \frac{7}{18} M_{(8)} + R^{\mu\nu} \left(\frac{5}{18} K_{\mu\nu(6)} - \frac{5}{6} L_{\mu\nu(4)} \right) \\
&+ \frac{20}{21} R^{\mu\alpha\nu\beta} K_{\mu\nu\alpha\beta(4)} - \frac{2}{5} \nabla^\mu R L_{\mu(4)} + \left(\frac{5}{3} K^{\mu\nu\alpha}{}_{(2)} \right. \\
&\left. + M^{\mu\nu\alpha} \right) K_{\mu\nu\alpha(4)} - \frac{12}{5} M^{\mu\nu\alpha} L_{\mu\nu\alpha(2)} + \frac{20}{9} K^{\mu\nu\alpha\beta\gamma} K_{\mu\nu\alpha\beta\gamma(2)} \\
&+ \frac{6}{25} K^{\mu\nu}{}_{(4)} K_{\mu\nu(4)} - \frac{27}{25} M^{\mu\nu}{}_{(2)} M_{\mu\nu(2)} + \frac{48}{25} K^{\mu\nu\alpha\beta}{}_{(2)} K_{\mu\nu\alpha\beta(2)} \\
&- \frac{18}{25} M^{\mu\nu\alpha\beta} M_{\mu\nu\alpha\beta} + \frac{16}{25} K^{\mu\nu\alpha\beta\gamma\rho} K_{\mu\nu\alpha\beta\gamma\rho} \\
&+ \left(\frac{101}{450} R^\mu_\alpha R^{\nu\alpha} + \frac{68}{525} R^\mu{}_{\alpha\beta\gamma} R^{\nu\alpha\beta\gamma} - \frac{1}{5} R^{\alpha\beta} R^\mu{}_\alpha{}^\nu{}_\beta \right) K_{\mu\nu(4)} \\
&+ \left(-\frac{2}{5} R^\mu_\alpha R^{\nu\alpha} - \frac{6}{25} R^\mu{}_{\alpha\beta\gamma} R^{\nu\alpha\beta\gamma} + \frac{6}{25} R^{\alpha\beta} R^\mu{}_\alpha{}^\nu{}_\beta \right) M_{\mu\nu(2)} \\
&+ \left(-\frac{1}{6} R^{\mu\nu} R^{\alpha\beta} - \frac{1}{3} R^{\lambda\alpha\nu\beta} \right) I_{(\alpha\beta)\mu\nu(2)} - \frac{1}{3} R^{\mu\nu} R^{\alpha\beta\gamma\delta} I_{\alpha\gamma\mu\nu\beta\delta} \\
&+ \left(-\frac{2}{5} R^{\mu\nu\alpha\beta} R^\sigma{}_\mu{}^\rho{}_\alpha - R^\beta_\lambda R^{\lambda\sigma\nu\rho} \right) L_{\nu\beta\sigma\rho} \\
&+ \left(-\frac{6}{5} R^\alpha_\rho R^{\rho\mu\beta\nu} + \frac{2588}{1575} R^\mu_\rho R^{\rho\alpha\nu\beta} + \frac{4}{25} R^{\mu\sigma\nu\rho} R^\alpha{}_\sigma{}^\beta{}_\rho \right. \\
&\left. + \frac{1048}{1575} R^{\mu\sigma\alpha\rho} R^\nu{}_\sigma{}^\beta{}_\rho + \frac{962}{1575} R^{\mu\alpha}{}_{\sigma\rho} R^{\nu\beta\sigma\rho} \right) K_{\mu\nu\alpha\beta(2)} \\
&+ \frac{1088}{1575} R^{\mu\alpha\beta\gamma} R^{\nu\sigma}{}_\beta{}^\rho K_{\mu\nu\alpha\gamma\sigma\rho} + R^{\mu\nu} \left\{ -\frac{4}{45} \nabla^\alpha R I_{\mu\nu\alpha(2)} \right. \\
&+ \left(\frac{7}{10} M_{\mu\alpha\beta} + \frac{5}{6} K_{\mu\alpha\beta(2)} \right) K_\nu{}^{\alpha\beta}{}_{(2)} - \frac{1}{3} K^{\sigma\rho}{}_{\mu(2)} I_{\sigma\rho\nu(2)}
\end{aligned}$$

$$
-\frac{2}{3}I_{\sigma\rho\mu\nu\alpha}K^{\sigma\rho}{}_{\alpha(2)} + \frac{5}{9}K_{\mu\alpha\beta\lambda\gamma}K_{\nu}{}^{\alpha\beta\lambda\gamma} - \frac{2}{3}I_{\sigma\rho\mu\alpha\beta}K^{\sigma\rho}{}_{\nu}{}^{\alpha\beta}
$$

$$
-\frac{4}{15}K_{\mu\nu\lambda\sigma\rho}M^{\lambda\sigma\rho}\Bigg\} + R^{\mu}{}_{(\alpha}{}^{\nu}{}_{\beta)}\Bigg\{\left(\frac{7}{5}K_{\mu\sigma\rho}{}^{\alpha\beta} - \frac{9}{10}K^{\alpha\beta}{}_{\mu\sigma\rho}\right.
$$

$$
\left.-\frac{16}{15}I^{\alpha\beta}{}_{\sigma\rho\mu}\right)M_{\nu}{}^{\sigma\rho} + \left(-\frac{7}{10}K_{\rho}{}^{\alpha\beta}{}_{(2)} - \frac{9}{10}K^{\alpha\beta}{}_{\rho(2)}\right.
$$

$$
\left.-\frac{4}{15}I^{\alpha\beta}{}_{\rho(2)}\right)M_{\mu\nu}{}^{\rho} - \frac{8}{45}\nabla^{\rho}RI_{\mu\nu\rho}{}^{\alpha\beta} - \frac{3}{10}\nabla^{\alpha}RK_{\mu\nu}{}^{\beta}{}_{(2)}
$$

$$
+\frac{20}{21}K_{\mu\rho}{}^{\alpha}{}_{(2)}K_{\nu}{}^{\rho\beta}{}_{(2)} + \frac{40}{21}K_{\mu}{}^{\rho\alpha\beta\sigma}K_{\nu\rho\sigma(2)}
$$

$$
+\frac{40}{21}K_{\mu\rho}{}^{\alpha\lambda\gamma}K_{\nu}{}^{\rho\beta}{}_{\lambda\gamma}\Bigg\}
$$

$$
-\frac{7}{450}R^{\mu}_{\nu}R^{\nu}_{\alpha}R^{\alpha}_{\beta}R^{\beta}_{\mu} + \frac{1}{90}R_{\mu\alpha}R^{\alpha}_{\nu}R^{\mu}{}_{\sigma}{}^{\nu}{}_{\rho}R^{\sigma\rho}
$$

$$
+\frac{817}{6300}R_{\mu\alpha}R^{\alpha}_{\nu}R^{\mu}{}_{\lambda\sigma\rho}R^{\nu\lambda\sigma\rho} + \frac{391}{4725}R^{\nu}_{\mu}R^{\beta}_{\alpha}R^{\mu}{}_{\sigma}{}^{\alpha}{}_{\rho}R^{\sigma}{}_{\nu}{}^{\rho}{}_{\beta}
$$

$$
-\frac{2243}{9450}R_{\mu\nu}R_{\alpha\beta}R^{\mu\alpha}{}_{\sigma\rho}R^{\nu\beta\sigma\rho} - \frac{1}{75}R_{\mu\nu}R^{\alpha\beta}R^{\mu}{}_{\sigma}{}^{\nu}{}_{\rho}R^{\sigma}{}_{\alpha}{}^{\rho}{}_{\beta}
$$

$$
-\frac{16}{4725}R^{\nu}_{\mu}R^{\mu}{}_{\sigma}{}^{\alpha}{}_{\rho}R^{\sigma}{}_{\lambda}{}^{\rho}{}_{\gamma}R^{\lambda}{}_{\nu}{}^{\gamma}{}_{\alpha} - \frac{7}{300}R_{\mu\nu}R^{\mu}{}_{\alpha}{}^{\nu}{}_{\beta}R^{\alpha}{}_{\lambda\sigma\rho}R^{\beta\lambda\sigma\rho}
$$

$$
+\frac{8}{675}R^{\nu}_{\mu}R^{\mu\lambda}{}_{\alpha\beta}R^{\alpha\beta}{}_{\sigma\rho}R^{\sigma\rho}{}_{\nu\lambda} + \frac{247}{9450}R^{\mu\nu}{}_{\alpha\beta}R^{\alpha\beta}{}_{\lambda\gamma}R^{\lambda\gamma}{}_{\sigma\rho}R^{\sigma\rho}{}_{\mu\nu}
$$

$$
-\frac{32}{4725}R^{\mu}{}_{\alpha}{}^{\nu}{}_{\beta}R^{\alpha}{}_{\lambda}{}^{\beta}{}_{\gamma}R^{\lambda}{}_{\rho}{}^{\gamma}{}_{\sigma}R^{\rho}{}_{\mu}{}^{\sigma}{}_{\nu} + \frac{1}{105}R^{\mu}{}_{\alpha\beta\gamma}R^{\nu\alpha\beta\gamma}R_{\mu\lambda\sigma\rho}R_{\nu}{}^{\lambda\sigma\rho}
$$

$$
+\frac{32}{4725}R_{\mu\alpha\nu\beta}R^{\alpha}{}_{\lambda}{}^{\beta}{}_{\gamma}R^{\mu}{}_{\sigma}{}^{\lambda}{}_{\rho}R^{\nu\sigma\gamma\rho} - \frac{232}{4725}R^{\mu\nu}{}_{\sigma\rho}R^{\alpha\beta\sigma\rho}R_{\mu\lambda\alpha\gamma}R^{\lambda}{}_{\nu}{}^{\gamma}{}_{\beta}
$$

$$
+\frac{299}{4725}R^{\mu\nu}{}_{\sigma\rho}R^{\alpha\beta\sigma\rho}R_{\mu\alpha\lambda\gamma}R_{\nu\beta}{}^{\lambda\gamma}\,. \qquad (2.165)
$$

This expression can be simplified a little bit to reduce the number of invariants (see [9, 11, 12]). Further transformations of the expression for the quantity $Z_{(6)}$, (2.163)–(2.165), in general form appear to be pointless since they are very cumbersome and there are very many independent invariants constructed from the curvatures and their covariant derivatives. The set of invariants and the form of the result should be chosen in accordance with the specific character of the considered problem. That is why we present the result for the De Witt coefficient $[a_4]$ by the set of formulas (2.129), (2.141)–(2.159) and (2.161)–(2.165) (an alternative reduced form is presented in [9, 11, 12, 4]. One should note that notation and the normalization of De Witt coefficients is slightly different in the cited references.

2.5 Effective Action of Massive Fields

Let us illustrate the elaborated methods for the calculation of the one-loop effective action on the example of real scalar, spinor and vector massive quantum matter fields on a classical gravitational field background in the four-dimensional physical space-time. In this section we follow our papers [6, 12].

The operator (1.22), (1.29) in this case has the form

$$\hat{\Delta} = \begin{cases} -\Box + \xi R + m^2, & (j=0)\,, \\ \gamma^\mu \nabla_\mu + m\,, & (j=1/2)\,, \\ -\delta d + m^2\,, & (j=1)\,, \end{cases} \tag{2.166}$$

where m and j are the mass and the spin of the field, ξ is the coupling constant of the scalar field with the gravitational field, γ^μ are the Dirac matrices, $\gamma^{(\mu}\gamma^{\nu)} = \hat{1}\, g^{\mu\nu}$, d is the exterior derivative and δ is the operator of co-differentiation on forms: $(\delta\varphi)_{[\mu_1\cdots\mu_k]} = \nabla^\mu \varphi_{[\mu\mu_1\cdots\mu_k]}$.

The commutator of covariant derivatives (3.13) (2.14) has the form

$$\mathcal{R}_{\mu\nu} = \begin{cases} 0 \\ \frac{1}{4}\gamma^\alpha\gamma^\beta R_{\alpha\beta\mu\nu}\,. \\ R^\alpha{}_{\beta\mu\nu} \end{cases} \tag{2.167}$$

Using the equations

$$\gamma^\mu\gamma^\nu\nabla_\mu\nabla_\nu = \hat{1}\left(\Box - \frac{1}{4}R\right)\,,$$

$$(d\delta + \delta d)\,\varphi^\mu = (\delta^\mu_\nu \Box - R^\mu_\nu)\,\varphi^\nu\,,$$

$$d^2 = \delta^2 = 0\,, \tag{2.168}$$

one can express the Green functions of the operator $\hat{\Delta}$, (2.166), in terms of the Green functions of the minimal operator, (1.30),

$$\hat{\Delta}^{-1} = \Pi\left(-\Box - Q + m^2\right)^{-1} , \tag{2.169}$$

where

$$\Pi = \begin{cases} 1 \\ m - \gamma^\mu \nabla_\mu , \\ 1 - \frac{1}{m^2} d\delta \end{cases} \qquad Q = \begin{cases} -\xi R \\ -\frac{1}{4}\hat{1}\, R\, . \\ -R^\alpha{}_\beta \end{cases} \tag{2.170}$$

Using the Schwinger–De Witt representation, (1.33),

$$\Gamma_{(1)} = -\frac{1}{2\mathrm{i}} \log \operatorname{sdet} \hat{\Delta} = \frac{1}{2\mathrm{i}} \int\limits_0^\infty \frac{\mathrm{d}s}{s} \operatorname{sdet} \exp(\mathrm{i}s\hat{\Delta}) , \tag{2.171}$$

and the equations

$$\begin{aligned} \operatorname{tr} \exp(\mathrm{i}s\gamma^\mu \nabla_\mu) &= \operatorname{tr} \exp(-\mathrm{i}s\gamma^\mu \nabla_\mu) , \\ \exp(\mathrm{i}s\delta d) &= \exp\{\mathrm{i}s(d\delta + \delta d)\} - \exp(\mathrm{i}s d\delta) + 1 , \\ \operatorname{tr} \exp(\mathrm{i}s d\delta)\Big|_{j=1} &= \operatorname{tr} \exp(\mathrm{i}s\, \Box)\Big|_{j=0} , \end{aligned} \tag{2.172}$$

we obtain up to non-essential infinite contributions $\sim \delta(0)$,

$$\log \det(\gamma^\mu \nabla_\mu + m) = \frac{1}{2} \log \det \left\{ \hat{1} \left(-\Box + \frac{1}{4} R + m^2 \right) \right\} ,$$

$$\begin{aligned} \log \det(-\delta d + m^2) &= \log \det\left(-\Box\, \delta^\beta_\alpha + R^\beta_\alpha + m^2 \delta^\beta_\alpha\right) \\ &\quad - \log \det(-\Box + m^2)\Big|_{j=0} . \end{aligned} \tag{2.173}$$

Thus we have reduced the functional determinants of the operators (2.166) to the functional determinants of the minimal operators of the form (1.30). By making use of the formulas (2.171)–(2.173), (1.50) and (1.54) we obtain the asymptotic expansion of the one-loop effective action in the inverse powers of the mass

$$\Gamma_{(1)\mathrm{ren}} = \frac{1}{2(4\pi)^2} \sum_{k \geq 3} \frac{B_k}{k(k-1)(k-2) m^{2(k-2)}} , \tag{2.174}$$

where

$$B_k = \begin{cases} A_k \\ \frac{1}{2} A_k \\ A_k\Big|_{(j=1)} - A_k\Big|_{(j=0,\xi=0)} \end{cases} \tag{2.175}$$

and the coefficients A_k are given by (1.55).

Let us stress that the coefficients A_k, (1.55), for the spinor field contain a factor (-1) in addition to the usual trace over the spinor indices according to the definition of the supertrace (1.25).

Substituting the matrices $\mathcal{R}_{\mu\nu}$, (2.167), and Q, (2.170), in (2.160) and using (2.175) we obtain the first coefficient, B_3, in the asymptotic expansion (2.174)

$$B_3 = \int \mathrm{d}^4x\, g^{1/2}\Big\{ c_1 R\Box R + c_2 R_{\mu\nu}\Box R^{\mu\nu} + c_3 R^3 + c_4 R R_{\mu\nu}R^{\mu\nu}$$

$$+c_5 R R_{\mu\nu\alpha\beta}R^{\mu\nu\alpha\beta} + c_6 R^\mu_\nu R^\nu_\lambda R^\lambda_\mu + c_7 R^{\mu\nu}R_{\alpha\beta}R^{\alpha\ \beta}_{\ \mu\ \nu}$$

$$+c_8 R_{\mu\nu}R^\mu{}_{\alpha\beta\gamma}R^{\nu\alpha\beta\gamma} + c_9 R_{\mu\nu}{}^{\alpha\beta}R_{\alpha\beta}{}^{\sigma\rho}R_{\sigma\rho}{}^{\mu\nu}$$

$$+c_{10} R^{\mu\ \nu}_{\ \alpha\ \beta}R^{\alpha\ \beta}_{\ \sigma\ \rho}R^{\sigma\ \rho}_{\ \mu\ \nu}\Big\}\,, \tag{2.176}$$

where the coefficients c_i are given in the Table 2.1.

The renormalized effective action (2.174) can be used to obtain the renormalized matrix elements of the energy-momentum tensor of the quantum matter fields in the background gravitational field [120, 121]

$$\Big\langle \mathrm{out,vac}|\mathrm{T}_{\mu\nu}(x)|\mathrm{in,vac}\Big\rangle_{\mathrm{ren}} = -\hbar\, 2\, g^{-1/2}\frac{\delta\Gamma_{(1)\mathrm{ren}}}{\delta g^{\mu\nu}(x)} + O(\hbar^2)\,. \tag{2.177}$$

Such problems were intensively investigated (see, for example, [137, 42]). In particular, in the papers [120, 121] the vacuum polarization of the quantum fields in the background gravitational field of the black holes was investigated. In these papers an expression for the renormalized one-loop effective action was obtained that is similar to (2.174)–(2.176) but does not take into account the terms, that do not contribute to the effective vacuum energy-momentum tensor (2.177) when the background metric satisfies the vacuum Einstein equations, $R_{\mu\nu}=0$. Our result (2.176) is valid, however, in general case of arbitrary background space. Moreover, using the results of Sect. 2.4 for the De Witt coefficient $[a_4]$, one can calculate the coefficient A_4 and, therefore, the next term, B_4 in the asymptotic expansion of the effective action (2.174) of order $1/m^4$. The technique for the calculation of the De Witt coefficients developed in this section is very algorithmic and can be realized on computers (all the needed information is contained in the first three sections of the present chapter) (see [44]). In this case one can also calculate the next terms of the expansion (2.174).

However, the effective action functional $\Gamma_{(1)}$ is, in general, essentially nonlocal and contains an imaginary part. The asymptotic expansion in the inverse powers of the mass (2.174) does not reflect these properties. It describes

Table 2.1. B_3 **for matter fields**

	Scalar field	Spinor field	Vector field
c_1	$\frac{1}{2}\xi^2 - \frac{1}{5}\xi + \frac{1}{56}$	$-\frac{3}{280}$	$-\frac{27}{280}$
c_2	$\frac{1}{140}$	$\frac{1}{28}$	$\frac{9}{28}$
c_3	$\left(\frac{1}{6} - \xi\right)^3$	$\frac{1}{864}$	$-\frac{5}{72}$
c_4	$-\frac{1}{30}\left(\frac{1}{6} - \xi\right)$	$-\frac{1}{180}$	$\frac{31}{60}$
c_5	$\frac{1}{30}\left(\frac{1}{6} - \xi\right)$	$-\frac{7}{1440}$	$-\frac{1}{10}$
c_6	$-\frac{8}{945}$	$-\frac{25}{756}$	$-\frac{52}{63}$
c_7	$\frac{2}{315}$	$\frac{47}{1260}$	$-\frac{19}{105}$
c_8	$\frac{1}{1260}$	$\frac{19}{1260}$	$\frac{61}{140}$
c_9	$\frac{17}{7560}$	$\frac{29}{7560}$	$-\frac{67}{2520}$
c_{10}	$-\frac{1}{270}$	$-\frac{1}{108}$	$\frac{1}{18}$

well the effective action only in weak gravitational fields ($R \ll m^2$). In strong gravitational fields ($R \gg m^2$), as well as for the massless matter fields, the asymptotic expansion (2.174) becomes meaningless. In this case it is necessary either to sum up some leading (in some approximation) terms or to use from the very beginning the non-local methods for the Green function and the effective action.

3. Partial Summation of Schwinger–De Witt Expansion

3.1 Summation of Asymptotic Expansions

The solution of the wave equation in background fields, (1.32), by means of the proper time method, (1.33), turns out to be very convenient for investigation of many general problems of the quantum field theory, especially for the analysis of the ultraviolet behavior of Green functions, regularization and renormalization. However, in practical calculations of concrete effects one fails to use the proper time method directly and one is forced to use model non-covariant methods.

In order to use the advantages of the covariant proper time method it is necessary to sum up the asymptotic series (1.43) for the evolution function. In general case the exact summation is impossible. Therefore, one can try to carry out the partial summation, i.e., to single out the leading (in some approximation) terms and sum them up in the first line. On the one hand one can limit oneself to a given order in background fields and sum up all derivatives, on the other hand one can neglect the derivatives and sum up all powers of background fields.

In this way we come across a certain difficulty. The point is that the asymptotic series do not converge, in general. Therefore, in the paper [46] it is proposed to give up the Schwinger–De Witt representation (1.33), (1.49), and treat it only as an auxiliary tool for the separation of the ultraviolet divergences. It is stated there that the Schwinger–De Witt representation, (1.33), exists for a small class of spaces — only when the semi-classical solution is exact.

However, the divergence of the asymptotic series (1.43) does not mean at all that one must give up the Schwinger–De Witt representation (1.33). The point is, the $\Omega(s)$ is not analytical at the point $s = 0$. Therefore, it is natural that the direct summation of the power series in s, (1.43), leads to divergences. In spite of this one can get a certain useful information from the structure of the asymptotic (divergent) series.

Let us consider a physical quantity $G(\alpha)$ which is defined by an asymptotic expansion of the perturbation theory in a parameter α

$$G(\alpha) = \sum_{k\geq 0} c_k \alpha^k \,. \tag{3.1}$$

The radius of convergence of the series (3.1) is given by the expression [210]

$$R = \left(\lim_{k\to\infty} \sup |c_k|^{1/k} \right)^{-1} . \tag{3.2}$$

If $R \neq 0$ then in the disc $|\alpha| < R$ of the complex plane of the parameter α the series (3.1) converges and defines an analytical function. If the considered physical quantity $G(\alpha)$ is taken to be analytical function, then outside the disc of convergence of the series (3.1), $|\alpha| \geq R$, it should be defined by analytical continuation. The function $\tilde{G}(\alpha)$ obtained in this way certainly have singularities, the first one lying on the circle $|\alpha| = R$. The analytical continuation through the boundary of the disc of convergence is impossible if all the points of the boundary (i.e., the circle $|\alpha| = R$) are singular. In this case the physical quantity $G(\alpha)$ appears to be meaningless for $|\alpha| \geq R$. If $R = 0$, then the series (3.1) diverges for any α, i.e., the function $G(\alpha)$ is not analytic in the point $\alpha = 0$. In this case it is impossible to carry out the summation and the analytical continuation. Nevertheless, one can gain an impression of t he exact quantity $G(\alpha)$ by making use of the Borel procedure for summation of asymptotic (in general, divergent) series [192].

The idea consists in the following. One constructs a new series with better convergence properties which reproduces the initial series by an integral transform. Let us define the Borel function

$$B(z) = \sum_{k\geq 0} \frac{c_k}{\Gamma(\mu k + \nu)} z^k , \tag{3.3}$$

where μ and ν are some complex numbers ($\mathrm{Re}\,\mu$, $\mathrm{Re}\,\nu > 0$).

The radius of convergence $\tilde{R}$ of the series (3.3) equals

$$\tilde{R} = \left(\lim_{k\to\infty} \sup \left| \frac{c_k}{\Gamma(\mu k + \nu)} \right|^{1/k} \right)^{-1} . \tag{3.4}$$

Thus, when the coefficients c_k of the series (3.1) rise not faster than $\exp(Mk \log k)$, $M = \mathrm{const}$, then one can always choose μ in such way, $\mathrm{Re}\,\mu \geq M$, that the radius of convergence of the series (3.3) will be not equal to zero $\tilde{R} \neq 0$, i.e., the Borel function $B(z)$ will be analytical at the point $z = 0$. Outside the disc of convergence, $|z| \geq \tilde{R}$, the Borel function is defined by analytical continuation.

Let us define

$$\tilde{G}(\alpha) = \int_C \mathrm{d}t\, t^{\nu-1} \mathrm{e}^{-t} B(\alpha t^{\mu}) , \tag{3.5}$$

where the integration contour C starts at the zero point and goes to infinity in the right half-plane ($\mathrm{Re}\, t \to +\infty$). The asymptotic expansion of the function $\tilde{G}(\alpha)$ for $\alpha \to 0$ has the form (3.1). Therefore, the function $\tilde{G}(\alpha)$, (3.5), (which is called the Borel sum of the series (3.1)) can be considered as the true physical quantity $G(\alpha)$.

The analytical properties of the Borel function $B(z)$ determine the convergence properties of the initial series (3.1). So, if the initial series (3.1) has a finite radius of convergence $R \neq 0$, then from (3.4) it follows that the Borel series (3.3) has an infinite radius of convergence $\tilde{R} = \infty$ and, therefore, the Borel function $B(z)$ is an entire function (analytical in any finite part of the complex plane). In this case the function $\tilde{G}(\alpha)$, (3.5), is equal to the sum of the initial series (3.1) for $|\alpha| < R$ and determines its analytical continuation outside the disc of convergence $|\alpha| \geq R$. If the Borel function $B(z)$ has singularities in the finite part of the complex plane, i.e., $\tilde{R} < \infty$, then from (3.4) it follows that the series (3.1) has a radius of convergence equal to zero $R = 0$, and, therefore, the physical quantity $G(\alpha)$ is not analytic at the point $\alpha = 0$. At the same time there always exist a region in the complex plane of the variable α where the Borel sum $\tilde{G}(\alpha)$ is still well defined and can be used for the analytical continuation to physical values of α. In this way different integration contours will give different functions $\tilde{G}(\alpha)$. In this case one should choose the contour of integration from some additional physical assumptions concerning the analytical properties of the exact function $G(\alpha)$.

3.2 Covariant Methods for Investigation of Nonlocalities

The De Witt coefficients a_k have the background dimension L^{-2k}, where L is the length unit. Therefore, the standard Schwinger–De Witt expansion, (1.43), (1.52) and (1.54), is, in fact, an expansion in the background dimension [34, 35]. In a given order in the background dimension L^{-2k} both the powers of the background fields, R^k, as well as their derivatives, $\nabla^{2(k-1)}R$, are taken into account. In order to investigate the nonlocalities it is convenient to reconstruct the local Schwinger–De Witt expansion in such a way that the expansion is carried out in the background fields but their derivatives are taken into account exactly from the very beginning. Doing this one can preserve the manifest covariance by using the methods developed in the Chap. 2.

Let us introduce instead of the Green function $G^A{}_B(x,y)$ of the operator (1.30), whose upper index belongs to the tangent space in the point x and the lower one — in the point y, a three-point Green function $\mathcal{G}^{A'}_{B'}(x,y|x')$, that depends on some additional fixed point x',

$$\mathcal{G}(x,y|x') = \mathcal{P}^{-1}(x,x')\Delta^{-1/2}(x,x')G(x,y)\Delta^{-1/2}(y,x')\mathcal{P}(y,x') \, . \tag{3.6}$$

This Green function is scalar at the points x and y and a matrix at the point x'. In the following we will not exhibit the dependence of all quantities on the fixed point x'.

The equation for the Green function $\mathcal{G}(x,y)$, (3.6), has the form (1.32)

$$\left(F_x - \hat{1}\, m^2\right) \mathcal{G}(x, y) = -\hat{1}\, g^{-1/2}(x) \Delta^{-1}(x) \delta(x, y) , \tag{3.7}$$

where F_x is the operator (2.112) and $\hat{1} = \delta^A{}_B$.

Let us single out in the operator F_x the free part that is of zero order in the background fields. Using (2.120) we have

$$F_x = \hat{1}\, \bar{\Box}_x + \tilde{F}_x , \tag{3.8}$$

where

$$\bar{\Box}_x = g^{\mu'\nu'}(x') \bar{\nabla}^x_{\mu'} \bar{\nabla}^x_{\nu'} , \tag{3.9}$$

$$\tilde{F}_x = \hat{1}\, \tilde{X}^{\mu'\nu'}(x) \bar{\nabla}^x_{\mu'} \bar{\nabla}^x_{\nu'} + Y^{\mu'}(x) \bar{\nabla}^x_{\mu'} + Z(x) , \tag{3.10}$$

$$\tilde{X}^{\mu'\nu'} = X^{\mu'\nu'}(x) - g^{\mu'\nu'}(x') , \tag{3.11}$$

and the operators $\bar{\nabla}_{\mu'}$ and the quantities $X^{\mu'\nu'}$, $Y^{\mu'}$ and Z are defined by the formulas (2.41), (2.46), (2.122) and (2.123). The operator $\tilde{F}$, (3.10), is of the first order in the background fields and can be considered as a perturbation.

By introducing the free Green function $\mathcal{G}_0(x, y)$,

$$\left(-\bar{\Box}_x + m^2\right) \mathcal{G}_0(x, y) = \hat{1}\, \Delta^{-1}(x) g^{-1/2}(x) \delta(x, y) , \tag{3.12}$$

and writing the equation (3.7) in the integral form

$$\mathcal{G}(x, z) = \mathcal{G}_0(x, z) + \int \mathrm{d}^n y\, g^{1/2}(y) \Delta(y) \mathcal{G}_0(x, y) \tilde{F}_y \mathcal{G}(y, z) , \tag{3.13}$$

we obtain from (3.13) by means of direct iterations

$$\mathcal{G}(x, z) = \mathcal{G}_0(x, z) + \sum_{k \geq 1} \int \mathrm{d}^n y_1\, g^{1/2}(y_1) \Delta(y_1) \cdots \mathrm{d}^n y_k\, g^{1/2}(y_k) \Delta(y_k)$$

$$\times \mathcal{G}_0(x, y_1) \tilde{F}_{y_1} \mathcal{G}_0(y_1, y_2) \cdots \tilde{F}_{y_k} \mathcal{G}_0(y_k, z) . \tag{3.14}$$

Using the covariant Fourier integral (2.55) and the equations (2.56) and (2.57) we obtain from (3.12) and (3.14) the momentum representations for the free Green function

$$\mathcal{G}_0(x, y) = \hat{1} \int \frac{\mathrm{d}^n k^{\mu'}}{(2\pi)^n} g^{1/2}(x') \exp\left\{ \mathrm{i} k_{\mu'} \left(\sigma^{\mu'}(y) - \sigma^{\mu'}(x) \right) \right\} \frac{1}{m^2 + k^2} , \tag{3.15}$$

and for the full Green function

$$\mathcal{G}(x, y) = \int \frac{\mathrm{d}^n k^{\mu'}}{(2\pi)^n} g^{1/2}(x') \frac{\mathrm{d}^n p^{\mu'}}{(2\pi)^n} g^{1/2}(x')$$

$$\times \exp\left\{ \mathrm{i} p_{\mu'} \sigma^{\mu'}(y) - \mathrm{i} k_{\mu'} \sigma^{\mu'}(x) \right\} \mathcal{G}(k, p) , \tag{3.16}$$

where

$$\mathcal{G}(k,p) = (2\pi)^n g^{-1/2}(x')\frac{\delta(k^{\mu'} - p^{\mu'})}{m^2 + k^2} + \frac{\Pi(k,p)}{(m^2 + k^2)(m^2 + p^2)} , \tag{3.17}$$

$$\Pi(k,p) = \tilde{F}(k,p) + \sum_{i\geq 1} \int \frac{d^n q_1^{\mu'}}{(2\pi)^n} g^{1/2}(x') \cdots \frac{d^n q_i^{\mu'}}{(2\pi)^n} g^{1/2}(x')$$

$$\times \tilde{F}(k,q_1)\frac{1}{m^2 + q_1^2} \cdots \tilde{F}(q_{i-1},q_i)\frac{1}{m^2 + q_i^2}\tilde{F}(q_i,p) , \tag{3.18}$$

$$\tilde{F}(k,p) = -\hat{1}\, \tilde{X}^{\mu'\nu'}(k-p)p_{\mu'}p_{\nu'} - iY^{\mu'}(k-p)p_{\mu'} + Z(k-p) , \tag{3.19}$$

$$k^2 \equiv g_{\mu'\nu'}(x')k^{\mu'}k^{\nu'} ,$$

and $\tilde{X}^{\mu'\nu'}(q)$, $Y^{\mu'}(q)$ and $Z(q)$ are the covariant Fourier components, (2.55), of the coefficients $\tilde{X}^{\mu'\nu'}$, $Y^{\mu'}$ and Z, (3.11), (2.46), (2.122), (2.123).

The formulas (3.15)–(3.19) reproduce the covariant generalization of the usual diagrammatic technique. Therefore, one can apply well elaborated methods of the Feynman momentum integrals. The Fourier components of the coefficient functions $\tilde{X}^{\mu'\nu'}(q)$, $Y^{\mu'}(q)$ and $Z(q)$, can be expressed in terms of the Fourier components of the background fields, $R^{\mu}{}_{\nu\alpha\beta}$, $\mathcal{R}_{\mu\nu}$ and Q, using the formulas obtained in Chap. 2. As usual [50, 155, 193, 42] one should choose the contour of integration over k_0 in the momentum integrals (3.15)–(3.18). Different ways of integration correspond to different Green functions. For the causal (Feynman) Green function one should either assume $k^2 \to k^2 - i\varepsilon$ or go to the Euclidean sector of the space-time [50, 155, 193].

Similarly, one can construct the kernels of any non-local operators of general form, $f(\Box)$, where $f(z)$ is some function. In the zeroth approximation in background fields we obtain

$$f(\Box)(x,y) = \mathcal{P}(x)\Delta^{1/2}(x)\bar{f}(\Box)(x,y)\Delta^{1/2}(y)\mathcal{P}^{-1}(y) ,$$

$$\bar{f}(\Box)(x,y) = \int \frac{d^n k^{\mu'}}{(2\pi)^n} g^{1/2}(x') \exp\left\{ik_{\mu'}\left(\sigma^{\mu'}(y) - \sigma^{\mu'}(x)\right)\right\} f(-k^2) . \tag{3.20}$$

An important method for the investigation of the nonlocalities is the analysis of the De Witt coefficients a_k and the partial summation of the asymptotic series (1.43). In this case one should limit oneself to some order in background fields and sum up all derivatives of background fields. In order to get an effective expansion in background fields it is convenient to change a little the "diagrammatic" technique for the De Witt coefficients developed in Sect. 2.3. Although all the terms in the sum (2.117) have equal background dimension, L^{n-2k}, they are of different order in background fields. From the formula (2.117) it is not seen immediately what order in background fields has each term of the sum (2.117), i.e., a single diagram, since all the diagrams for the coefficient $< n|a_k >$ have k blocks. However, among these blocks, i.e., the matrix elements $< m|F|m+2 >$, there are dimensionless blocks that do not have any background dimension and are of zero order in background fields. These are the blocks (matrix elements $< m|F|m+2 >$) with the number of outgoing lines equal to the number of the incoming lines plus 2 (c.f. (2.124)). Therefore, one can order all the diagrams for the De Witt coefficients (i.e., different terms of the sum (2.117)) in the following way. The first diagram contains only one dimensional block, all others being dimensionless. The second class contains all diagrams with two dimensional blocks, the third one — three etc. The last diagram contains k dimensional blocks. To obtain the De Witt coefficients in the first order in background fields it is sufficient to restrict oneself to the first diagram. To get the De Witt coefficients in the second order in background fields it is sufficient to restrict oneself to the first diagram and the set of diagrams with two dimensional blocks etc. This method is completely analogous to the separation of the free part of the operator F, (3.8). The dimensional matrix elements $< m|F|n >$, (with $m \geq n$), of the operator F, (2.120), are equal to the matrix elements of the operator $\tilde{F}$, (3.10). When calculating the matrix elements (2.124) and (2.125) one can also neglect the terms that do not contribute in the given order in background fields.

After such reconstruction (and making use of (2.124)) the formula (2.117) for the De Witt coefficients $< n|a_k >$ takes the form

$$
\begin{aligned}
< n|a_k >= & \sum_{1\leq N\leq k;\ 1\leq i_1<i_2<\cdots<i_{N-1}\leq k-1;\ n_i} \ \prod_{1\leq j\leq N} \frac{\binom{2i_j+n_j-1}{i_{j-1}}}{\binom{2i_j+n_j-1}{i_j}} \\
& \times < n; k-i_{N-1}-1|F|n_{N-1} >< n_{N-1}; i_{N-1}-i_{N-2}-1|F|n_{N-2} > \\
& \times \cdots < n_2; i_2-i_1-1|F|n_1 >< n_1; i_1-1|F|0 > ,
\end{aligned}
\qquad (3.21)
$$

where the following notation is introduced

$$
< n;k|F|m > \equiv g^{\nu_1\nu_2}\cdots g^{\nu_{2k-1}\nu_{2k}} < \nu_1\cdots\nu_{n+2k}|F|\mu_1\cdots\mu_m > , \qquad (3.22)
$$

$$
i_0 \equiv 0, \qquad i_N \equiv k, \qquad n_N \equiv n,
$$

and the summation over n_i is carried out in such limits that all matrix elements are dimensional, (i.e., for each $< n;k|F|m >$, (3.22), there holds: $n+2k \geq m$):

$$
n_1+2(i_1-1) \geq 0\ ,\ \ n_2+2(i_2-i_1-1) \geq n_1\ ,\ \ \cdots\ ,\ \ n+2(k-i_{N-1}-1) \geq n_{N-1}\ .
$$

The formula (3.21) also enables one to use the diagrammatic technique. However, for analyzing the general structure of the De Witt coefficients and for the partial summation it is no longer effective. Therefore, one should use the analytic expression (3.21).

3.3 Summation of First-Order Terms

Let us calculate the coincidence limit of the De Witt coefficients $[a_k] =< 0|a_k >$ in the first order in background fields. Using the formula (3.21) we obtain

$$[a_k] = \frac{1}{\binom{2k-1}{k}} < 0; k-1|F|0> + O(R^2)\,, \qquad (3.23)$$

where $O(R^2)$ denotes all omitted terms of second order in background fields.

Using (3.22), (2.121)–(2.125) and (2.46) and the formulas of the Sect. 2.2 we obtain from (3.23) up to quadratic terms

$$[a_k] = \frac{k!(k-1)!}{(2k-1)!}\,\Box^{k-1}\left(Q + \frac{k}{2(2k+1)}\hat{1}\,R\right) + O(R^2)\,, \qquad (k \geq 1)\,. \quad (3.24)$$

The expression (3.24) can be used for the calculation of the $\Omega(s|x,x)$ in the first order in background fields. Substituting (3.24) in (1.43), we obtain

$$\Omega(s|x,x) = \hat{1} + \mathrm{i}s\left(f_1(\mathrm{i}s\,\Box)Q + \hat{1}\,f_2(\mathrm{i}s\,\Box)R\right) + O(R^2)\,, \qquad (3.25)$$

where

$$f_1(z) = \sum_{k\geq 0}\frac{k!}{(2k+1)!}z^k, \qquad (3.26)$$

$$f_2(z) = \sum_{k\geq 0}\frac{(k+1)!(k+1)}{(2k+3)!}z^k. \qquad (3.27)$$

The power series (3.26) and (3.27) converge for any finite z and hence define entire functions. One can sum up the series of the type (3.26) and (3.27) using the general formula

$$\frac{(k+l)!}{(2k+2l+1)!} = \frac{l!}{(2l)!}\int\limits_0^1 \mathrm{d}\xi\,\xi^{2l}\frac{1}{k!}\left(\frac{1-\xi^2}{4}\right)^k, \qquad (3.28)$$

that is easily obtained from the definition of the Euler beta-function [98].

Substituting (3.28) in (3.26) and (3.27) and summing over k we obtain the integral representations of the functions $f_1(z)$ and $f_2(z)$

$$f_1(z) = \int\limits_0^1 \mathrm{d}\xi\,\exp\left\{\frac{1}{4}\left(1-\xi^2\right)z\right\}\,, \qquad (3.29)$$

$$f_2(z) = \int\limits_0^1 \mathrm{d}\xi\,\frac{1}{4}(1-\xi^2)\exp\left\{\frac{1}{4}\left(1-\xi^2\right)z\right\}\,. \qquad (3.30)$$

The kernels of the non-local operators $f_1(\mathrm{i}s\,\Box)$, $f_2(\mathrm{i}s\,\Box)$ should be understood in terms of covariant momentum expansions (3.20).

Using the obtained , (3.25), one can easily obtain the Green function at coinciding points, $G(x,x)$, in the first order in background fields. Substituting (3.25) in (1.49) and supposing $\mathrm{Im}\, m^2 < 0$ (for the causal Green function), we obtain after the integration over the proper time in the n-dimensional space

$$G(x,x) = \mathrm{i}(4\pi)^{-n/2}\Bigg\{\Gamma\left(1-\frac{n}{2}\right)\hat{1}\, m^{n-2} \\ +\Gamma\left(2-\frac{n}{2}\right)m^{n-4}\left[\tilde{F}_1\left(-\frac{\Box}{4m^2}\right)Q+\hat{1}\,\tilde{F}_2\left(-\frac{\Box}{4m^2}\right)R\right]\Bigg\} \\ +O(R^2)\,, \tag{3.31}$$

where $\Gamma(z)$ is the Euler gamma-function, and

$$\tilde{F}_1(z) = \int\limits_0^1 \mathrm{d}\xi\ \left[1+(1-\xi^2)z\right]^{(n-4)/2}\,, \tag{3.32}$$

$$\tilde{F}_2(z) = \int\limits_0^1 \mathrm{d}\xi\,\frac{1}{4}(1-\xi^2)\left[1+(1-\xi^2)z\right]^{(n-4)/2}\,. \tag{3.33}$$

By expanding in the dimension n in the neighborhood of the point $n=4$ and subtracting the pole $1/(n-4)$ we obtain the renormalized Green function, $G_{\mathrm{ren}}(x,x)$, in the physical four-dimensional space-time (1.51), (1.52), up to terms of second order in background fields

$$G_{\mathrm{ren}}(x,x) = \frac{\mathrm{i}}{(4\pi)^2}\left\{F_1\left(-\frac{\Box}{4m^2}\right)Q+\hat{1}\,F_2\left(-\frac{\Box}{4m^2}\right)R\right)\Bigg\}+O(R^2)\,, \tag{3.34}$$

where

$$F_1(z) = 2 - J(z)\,, \tag{3.35}$$

$$F_2(z) = \frac{1}{18}\left(5-\frac{3}{z}\right)-\frac{1}{6}\left(1-\frac{1}{2z}\right)J(z)\,, \tag{3.36}$$

$$J(z) = 2(1+z)\int\limits_0^1 \frac{\mathrm{d}\xi}{1+(1-\xi^2)z}\,. \tag{3.37}$$

The formfactors $F_1(z)$, (3.35), and $F_2(z)$, (3.36), are normalized according the conditions

$$F_1(0) = F_2(0) = 0\,. \tag{3.38}$$

The integral (3.37) determines an analytical single-valued function in the complex plane z with a cut along the negative part of the real axis from -1 to $-\infty$:

$$J(z) = 2\sqrt{1+\frac{1}{z}}\log\left(\sqrt{z+1}+\sqrt{z}\right), \qquad |\arg(z+1)| < \pi, \tag{3.39}$$

$$J(x \pm \mathrm{i}\varepsilon) = 2\sqrt{1+\frac{1}{x}}\log\left(\sqrt{-x-1}+\sqrt{-x}\right) \pm \mathrm{i}\pi\sqrt{1+\frac{1}{x}}, \qquad (x < -1). \tag{3.40}$$

Thus we obtained the non-local expression for the Green function at coinciding points, (3.34). It reproduces the local Schwinger–De Witt expansion, (1.52), up to quadratic terms in eternal fields by expanding in inverse powers of the mass. The power series determining the formfactors $F_1(z)$, (3.35), and $F_2(z)$, (3.36), converge in the region $|z| < 1$. For $z = -1$ there is a threshold singularity – the branching point. Outside the circle $|z| \geq 1$ the formfactors are defined by the analytical continuation. The boundary conditions for the formfactors fix uniquely the ambiguity in the Green function (3.34). For the causal Green function ($\mathrm{Im}\, m^2 < 0$) the lower bank of the cut is the physical one. Therefore, with account of (3.40), the imaginary parts of the formfactors (3.35) and (3.36) in the pseudo-Euclidean region above the threshold $z = x - \mathrm{i}\varepsilon$, $x < -1$, equal

$$\mathrm{Im} F_1(x - \mathrm{i}\varepsilon) = \pi\sqrt{1+\frac{1}{x}}, \tag{3.41}$$

$$\mathrm{Im} F_2(x - \mathrm{i}\varepsilon) = \frac{\pi}{6}\left(1 - \frac{1}{2x}\right)\sqrt{1+\frac{1}{x}}. \tag{3.42}$$

The ultraviolet asymptotics $|z| \to \infty$ of the formfactors (3.35) and (3.36) have the form

$$F_1(z)\Big|_{|z|\to\infty} = -\log(4z) + 2 + O\left(\frac{1}{z}\log z\right), \tag{3.43}$$

$$F_2(z)\Big|_{|z|\to\infty} = \frac{1}{6}\left(-\log(4z) + \frac{5}{3}\right) + O\left(\frac{1}{z}\log z\right). \tag{3.44}$$

Let us consider the case of the massless field, $m^2 = 0$. Taking the limit $m^2 \to 0$ in (3.31) for $\mathrm{Re}\, n > 2$ we obtain the Green function of the massless field at coinciding points in the n-dimensional space in the first order in background fields

$$G(x,x) = \mathrm{i}(4\pi)^{-n/2}\frac{\Gamma\left(2-\frac{n}{2}\right)\left(\Gamma\left(\frac{n}{2}-1\right)\right)^2}{\Gamma(n-2)}$$

$$\times(-\Box)^{(n-4)/2}\left(Q + \frac{n-2}{4(n-1)}\hat{1}\, R\right) + O(R^2). \tag{3.46}$$

The formula (3.46) determines the analytic function of the dimension n in the region $2 < \mathrm{Re}\, n < 4$. After the analytical continuation there appear poles in the points $n = 4, 6, 8, \ldots$, that reflect the ultraviolet divergences, and a pole in the point $n = 2$ reflecting the infrared divergence. In the two-dimensional space (3.46) gives

$$G(x,x) = \frac{\mathrm{i}}{4\pi}\left\{-2\left(\frac{2}{n-2} + \mathbf{C} + \log\frac{-\Box}{4\pi\mu^2}\right)\frac{1}{\Box}Q - \hat{1}\,\frac{1}{\Box}R\right\} + O(R^2)\,. \tag{3.47}$$

For even dimensions, $n = 2N$, $(N \geq 2)$, from (3.46) we obtain

$$G(x,x) = \frac{\mathrm{i}}{(4\pi)^N}\cdot\frac{(N-2)!}{(2N-3)!}\left\{-\left(\frac{2}{n-2N} + \Psi(N-1)\right.\right.$$

$$\left. -2\Psi(2N-2) + \log\frac{-\Box}{4\pi\mu^2}\right)\Box^{N-2}\left(Q + \frac{N-1}{2(2N-1)}\hat{1}\,R\right)$$

$$\left. -\frac{1}{2(2N-1)^2}\Box^{N-2}R\right\} + O(R^2)\,, \tag{3.48}$$

where

$$\Psi(k) = -\mathbf{C} + \sum_{1\leq l\leq k-1}\frac{1}{l}\,. \tag{3.49}$$

In particular, in the physical four-dimensional space-time, $n \to 4$, we have from (3.48)

$$G(x,x) = \frac{\mathrm{i}}{(4\pi)^2}\left\{-\left(\frac{2}{n-4} + \mathbf{C} - 2 + \log\frac{-\Box}{4\pi\mu^2}\right)\left(Q + \frac{1}{6}\hat{1}\,R\right)\right.$$

$$\left. -\frac{1}{18}\hat{1}\,R\right\} + O(R^2)\,. \tag{3.50}$$

The renormalized Green function of the massless field can be obtained by using the ultraviolet asymptotics of the formfactors (3.43) and (3.44) and substituting the renormalization parameter instead of the mass, $m^2 \to \mu^2$. This reduces simply to a change of the normalization of the formfactors (3.38).

Let us stress that in the massless case the divergence of the coincidence limit of the Green function in the first order in background fields, (3.48), (3.50), is proportional to the De Witt coefficient $[a_{(n-2)/2}]$, (3.24). Therefore, in the conformally invariant case [42, 137],

$$Q = -\frac{n-2}{4(n-1)}\hat{1}\,R\,, \tag{3.51}$$

the linear part of the coefficient $[a_{(n-2)/2}]$, (3.24), is equal to zero and the Green function at coinciding points is finite in the first order in background fields, (3.47)–(3.50). In odd dimensions, $n = 2N + 1$, the Green function, (3.31), (3.46), is finite.

3.4 Summation of Second-Order Terms

Let us calculate the coincidence limit of the De Witt coefficients $[a_k] =< 0|a_k >$ in the second order in background fields. From the formula (3.21) we have up to cubic terms in background fields

$$[a_k] = \frac{1}{\binom{2k-1}{k}} < 0; k-1|F|0 >$$

$$+ \sum_{1\le i\le k-1;\ 0\le n_i \le 2(k-i-1)} \sum \frac{\binom{2k-1}{i}}{\binom{2k-1}{k}\binom{2i+n_i-1}{i}}$$

$$\times < 0; k-i-1|F|n_i >< n_i; i-1|F|0 > +O(R^3) \,, \qquad (3.52)$$

where $O(R^3)$ denotes all omitted terms of third order in background fields.

The total number of terms, that are quadratic in background fields and contain arbitrary number of derivatives, is infinite. Therefore, we will also neglect the total derivatives and the trace-free terms in (3.52) assuming to use the results for the calculation of the coefficients A_k, (1.55), and the one-loop effective action, (1.50), (1.54).

The number of the terms quadratic in background fields, in the coefficients A_k is finite. Let us write the general form of the coefficients A_k up to the terms of the third order in background fields

$$A_k = \int \mathrm{d}^n x\, g^{1/2} \mathrm{str}\, \frac{1}{2}\Bigg\{ \alpha_k Q \,\Box^{k-2} Q - 2\beta_k J_\mu \,\Box^{k-3} J^\mu + \gamma_k Q \,\Box^{k-2} R$$

$$+\hat{1}\left(\delta_k R_{\mu\nu} \,\Box^{k-2} R^{\mu\nu} + \varepsilon_k R \,\Box^{k-2} R\right) + O(R^3)\Bigg\} \,, \qquad (3.53)$$

where $J_\mu = \nabla_\alpha \mathcal{R}^\alpha{}_\mu$.

All other quadratic invariants can be reduced to those written above up to third order terms using the integration by parts and the Bianci identity. For example,

$$\int \mathrm{d}^n x\, g^{1/2} \mathrm{str}\, \mathcal{R}_{\mu\nu} \,\Box^{k} \mathcal{R}^{\mu\nu} = -2 \int \mathrm{d}^n x\, g^{1/2} \mathrm{str}\, \left\{ J_\mu \,\Box^{k-1} J^\mu + O(R^3) \right\} \,,$$

$$\int \mathrm{d}^n x\, g^{1/2} R_{\mu\nu\alpha\beta} \Box^k R^{\mu\nu\alpha\beta} = \int \mathrm{d}^n x\, g^{1/2} \Big\{ 4R_{\mu\nu} \Box^k R^{\mu\nu} - R \Box^k R + O(R^3) \Big\} . \tag{3.54}$$

Using (3.22), (2.121)–(2.125) and (2.46) and the formulas of the Sect. 2.2 one can calculate the coefficients α_k, β_k, γ_k, δ_k and ε_k from (3.52) and (1.55). Omitting the intermediate computations we present the result

$$\alpha_k = \frac{k!(k-2)!}{(2k-3)!} , \tag{3.55}$$

$$\beta_k = \frac{k!(k-1)!}{(2k-1)!} , \tag{3.56}$$

$$\gamma_k = 2(k-1)\frac{k!(k-1)!}{(2k-1)!} , \tag{3.57}$$

$$\delta_k = 2\frac{k!k!}{(2k+1)!} , \tag{3.58}$$

$$\varepsilon_k = (k^2-k-1)\frac{k!k!}{(2k+1)!} . \tag{3.59}$$

These coefficients were computed completely independently in the papers [8, 10, 12, 52].

Using the obtained coefficients, (3.55)–(3.59), we calculate the trace of the $\Omega(s)$, (1.43),

$$\int \mathrm{d}^n x\, g^{1/2} \mathrm{str}\, \Omega(s|x,x) = \sum_{k\ge 0} \frac{(\mathrm{i}s)^k}{k!} A_k . \tag{3.60}$$

Substituting (3.53) and (3.55)–(3.59) in (3.60) we obtain

$$\begin{aligned}\int \mathrm{d}^n x\, g^{1/2} \mathrm{str}\, \Omega(s|x,x) = \int \mathrm{d}^n x\, g^{1/2} \mathrm{str} \Bigg\{ \hat{1} + \mathrm{i}s \left(Q + \frac{1}{6}\hat{1}\, R \right) \\ + \frac{(\mathrm{i}s)^2}{2} \Bigg[Q f_1(\mathrm{i}s\,\Box) Q + 2J_\mu f_3(\mathrm{i}s\,\Box) \frac{1}{(-\Box)} J^\mu + 2Q f_2(\mathrm{i}s\,\Box) R \\ + \hat{1} \left(R_{\mu\nu} f_4(\mathrm{i}s\,\Box) R^{\mu\nu} + R f_5(\mathrm{i}s\,\Box) R \right) \Bigg] + O(R^3) \Bigg\} ,\end{aligned} \tag{3.61}$$

where

$$f_3(z) = \sum_{k\ge 0} \frac{(k+1)!}{(2k+3)!} z^k , \tag{3.62}$$

$$f_4(z) = \sum_{k\geq 0} 2\frac{(k+2)!}{(2k+5)!} z^k, \tag{3.63}$$

$$f_5(z) = \sum_{k\geq 0} \frac{(k+2)!}{(2k+5)!}(k^2+3k+1) z^k, \tag{3.64}$$

and $f_1(z)$ and $f_2(z)$ are given by the formulas (3.26), (3.27), (3.29) and (3.30).

The series (3.62)–(3.64) converge for any finite z and define in the same manner as (3.26) and (3.27) entire functions. The summation of the series (3.62) and (3.63) can be performed by means of the formula (3.28). Substituting (3.28) in (3.62) and (3.63) and summing over k we obtain the integral representation of the functions $f_3(z)$ and $f_4(z)$:

$$f_3(z) = \int\limits_0^1 d\xi \, \frac{1}{2}\xi^2 \exp\left\{\frac{1}{4}(1-\xi^2)z\right\}, \tag{3.65}$$

$$f_4(z) = \int\limits_0^1 d\xi \, \frac{1}{6}\xi^4 \exp\left\{\frac{1}{4}(1-\xi^2)z\right\}. \tag{3.65}$$

Noting that from (3.64) it follows

$$f_5(z) = \frac{1}{16} f_1(z) - \frac{1}{4} f_3(z) - \frac{1}{8} f_4(z), \tag{3.67}$$

we obtain from the formulas (3.29), (3.65) and (3.66)

$$f_5(z) = \int\limits_0^1 d\xi \, \frac{1}{8}\left(\frac{1}{2} - \xi^2 - \frac{1}{6}\xi^4\right) \exp\left\{\frac{1}{4}(1-\xi^2)z\right\}. \tag{3.68}$$

Using (3.61) one can calculate the one-loop effective action up to cubic terms in background fields. Substituting (3.61) in (1.50) and assuming $\operatorname{Im} m^2 < 0$, after integration over the proper time we obtain in the n-dimensional space

$$\begin{aligned}
\Gamma_{(1)} = \frac{1}{2(4\pi)^{n/2}} \int d^n x \, g^{1/2} \mathrm{str} \Bigg\{ & \Gamma\left(-\frac{n}{2}\right) \hat{1}\, m^n + \Gamma\left(1-\frac{n}{2}\right) m^{n-2} \left(Q + \frac{1}{6}\hat{1}\, R\right) \\
& + \frac{1}{2}\Gamma\left(2-\frac{n}{2}\right) m^{n-4} \Bigg[Q \tilde{F}_1\left(-\frac{\Box}{4m^2}\right) Q \\
& + 2J_\mu \frac{1}{(-\Box)} \tilde{F}_3\left(-\frac{\Box}{4m^2}\right) J^\mu + 2Q\tilde{F}_2\left(-\frac{\Box}{4m^2}\right) R \\
& + \hat{1}\left(R_{\mu\nu}\tilde{F}_4\left(-\frac{\Box}{4m^2}\right) R^{\mu\nu} + R\tilde{F}_5\left(-\frac{\Box}{4m^2}\right) R\right)\Bigg] + O(R^3)\Bigg\},
\end{aligned} \tag{3.69}$$

where

$$\tilde{F}_3(z) = \int\limits_0^1 \mathrm{d}\xi \, \frac{1}{2}\xi^2 \left[1 + (1 - \xi^2) z\right]^{(n-4)/2} , \tag{3.70}$$

$$\tilde{F}_4(z) = \int\limits_0^1 \mathrm{d}\xi \, \frac{1}{6}\xi^4 \left[1 + (1 - \xi^2) z\right]^{(n-4)/2} , \tag{3.71}$$

$$\tilde{F}_5(z) = \int\limits_0^1 \mathrm{d}\xi \, \frac{1}{8}\left(\frac{1}{2} - \xi^2 - \frac{1}{6}\xi^4\right) \left[1 + (1 - \xi^2) z\right]^{(n-4)/2} , \tag{3.72}$$

and $\tilde{F}_1(z)$ and $\tilde{F}_2(z)$ are given by the formulas (3.32) and (3.33).

Subtracting the pole in the dimension $1/(n-4)$, we obtain the renormalized effective action in the physical four-dimensional space-time, (1.53), (1.54), up to terms of third order in background fields

$$\begin{aligned}
\Gamma_{(1)\mathrm{ren}} = \frac{1}{2(4\pi)^2} \int \mathrm{d}^4x \, g^{1/2} \mathrm{str} \frac{1}{2} \Bigg\{ & Q F_1\left(-\frac{\Box}{4m^2}\right) Q \\
& + 2 J_\mu \frac{1}{(-\Box)} \tilde{F}_3\left(-\frac{\Box}{4m^2}\right) J^\mu + 2 Q \tilde{F}_2\left(-\frac{\Box}{4m^2}\right) R \\
& + \hat{1}\left(R_{\mu\nu} \tilde{F}_4\left(-\frac{\Box}{4m^2}\right) R^{\mu\nu} + R \tilde{F}_5\left(-\frac{\Box}{4m^2}\right) R\right)\Bigg] + O(R^3)\Bigg\} ,
\end{aligned} \tag{3.73}$$

where

$$F_3(z) = \frac{4}{9} + \frac{1}{3z} - \frac{1}{6}\left(1 + \frac{1}{z}\right) J(z) , \tag{3.74}$$

$$F_4(z) = \frac{23}{225} + \frac{7}{45z} + \frac{1}{15z^2} - \frac{1}{30}\left(1 + \frac{1}{z}\right)^2 J(z) , \tag{3.75}$$

$$F_5(z) = \frac{1}{900} - \frac{37}{360z} - \frac{1}{120z^2} - \frac{1}{60}\left(1 - \frac{3}{z} - \frac{1}{4z^2}\right) J(z) , \tag{3.76}$$

and $F_1(z)$, $F_2(z)$ and $J(z)$ are given by the formulas (3.35)–(3.37).

The formfactors $F_3(z)$, (3.74), $F_4(z)$, (3.75), and $F_5(z)$, (3.76), are normalized by the conditions

$$F_3(0) = F_4(0) = F_5(0) = 0 . \tag{3.77}$$

The normalization conditions of the formfactors (3.38) and (3.77) correspond to the normalization of the effective action

$$\Gamma_{(1)\text{ren}}\Big|_{m^2\to\infty} = 0 \,. \tag{3.78}$$

Thus, by means of the partial summation of the local Schwinger–De Witt expansion, (1.43), (3.60), we obtained a non-local expression for the effective action up to terms of third order in background fields, (3.73). Although the power series, that define the formfactors, i.e., the power series for the function $J(z)$, (3.37), converge only in the region $|z| < 1$, $z = -1$ being the threshold branching point, the expressions (3.35)–(3.37) and (3.74)–(3.76) are valid for any z. That means that the proper time method automatically does the analytical continuation in the ultraviolet region $|z| \to \infty$. All the ambiguity, which arises by the partial summation of the asymptotic expansion (3.60), reduces to the arbitrariness in the boundary conditions for the formfactors. Specifying the causal boundary conditions leads to the single-valued expression for the effective action. Using the prescription $m^2 \to m^2 - \mathrm{i}\varepsilon$ and the equation (3.40) we obtain the imaginary parts of the formfactors (3.74)–(3.76) in the pseudo-Euclidean region ($z = x - \mathrm{i}\varepsilon$) above the threshold ($x < -1$)

$$\operatorname{Im} F_3(x - \mathrm{i}\varepsilon) = \frac{\pi}{6}\left(1 + \frac{1}{x}\right)\sqrt{1 + \frac{1}{x}} \,, \tag{3.79}$$

$$\operatorname{Im} F_4(x - \mathrm{i}\varepsilon) = \frac{\pi}{30}\left(1 + \frac{1}{x}\right)^2\sqrt{1 + \frac{1}{x}} \,, \tag{3.80}$$

$$\operatorname{Im} F_5(x - \mathrm{i}\varepsilon) = \frac{\pi}{60}\left(1 - \frac{3}{x} - \frac{1}{4x^2}\right)\sqrt{1 + \frac{1}{x}} \,. \tag{3.81}$$

The imaginary parts of all formfactors, (3.41), (3.42), (3.79)–(3.81), are positive. This ensures the fulfillment of the important condition

$$\operatorname{Im} \Gamma_{(1)\text{ren}} > 0 \,. \tag{3.82}$$

The ultraviolet asymptotics $|z| \to \infty$ of the formfactors (3.74)–(3.76) have the form

$$F_3(z)\Big|_{|z|\to\infty} = \frac{1}{6}\left(-\log(4z) + \frac{8}{3}\right) + O\left(\frac{1}{z}\log z\right) \,, \tag{3.83}$$

$$F_4(z)\Big|_{|z|\to\infty} = \frac{1}{30}\left(-\log(4z) + \frac{46}{15}\right) + O\left(\frac{1}{z}\log z\right) \,, \tag{3.84}$$

$$F_5(z)\Big|_{|z|\to\infty} = \frac{1}{60}\left(-\log(4z) + \frac{1}{15}\right) + O\left(\frac{1}{z}\log z\right) \,. \tag{3.85}$$

Let us consider the case of the massless field. Taking the limit $m^2 \to 0$ in (3.69) in the region $\operatorname{Re} n > 2$ we obtain the one-loop effective action for the massless field in the n-dimensional space up to terms of third order in background fields

$$\Gamma_{(1)} = \frac{1}{2(4\pi)^{n/2}} \int d^n x\, g^{1/2} \frac{\Gamma\left(2-\frac{n}{2}\right)\left(\Gamma\left(\frac{n}{2}-1\right)\right)^2}{\Gamma(n-2)}$$

$$\times \operatorname{str}\left\{ \frac{1}{2} Q \left(-\Box\right)^{(n-4)/2} Q + \frac{1}{2(n-1)} J_\mu \left(-\Box\right)^{(n-6)/2} J^\mu \right.$$

$$+ \frac{n-2}{4(n-1)} Q \left(-\Box\right)^{(n-4)/2} R$$

$$+ \frac{1}{4(n^2-1)} \hat{1} \left[R_{\mu\nu} \left(-\Box\right)^{(n-4)/2} R^{\mu\nu} \right.$$

$$\left. \left. + \frac{1}{8}(n^2 - 2n - 4) R \left(-\Box\right)^{(n-4)/2} R \right] + O(R^3) \right\} . \qquad (3.86)$$

The formula (3.86) defines an analytical function of the dimension n in the region $2 < \mathrm{Re}\, n < 4$. The analytical continuation leads to the poles in the points $n = 2, 4, 6, \ldots$. In odd dimensions the expression (3.86) is finite and directly defines the effective action. In even dimensions the expression (3.86) is proportional to the De Witt coefficient $A_{n/2}$, (3.53).

Separating the pole $1/(n-4)$ in (3.86) we obtain the effective action for the massless field in the physical four-dimensional space-time up to terms $O(R^3)$

$$\Gamma_{(1)} = \frac{1}{2(4\pi)^2} \int d^n x\, g^{1/2} \operatorname{str} \left\{ \frac{1}{2} Q \left(-\frac{2}{n-4} - \mathbf{C} + 2 - \log \frac{-\Box}{4\pi\mu^2} \right) Q \right.$$

$$+ \frac{1}{6} J_\mu \frac{1}{(-\Box)} \left(-\frac{2}{n-4} - \mathbf{C} + \frac{8}{3} - \log \frac{-\Box}{4\pi\mu^2} \right) J^\mu$$

$$+ \frac{1}{6} Q \left(-\frac{2}{n-4} - \mathbf{C} + \frac{5}{3} - \log \frac{-\Box}{4\pi\mu^2} \right) R$$

$$+ \hat{1} \left[\frac{1}{60} R_{\mu\nu} \left(-\frac{2}{n-4} - \mathbf{C} + \frac{46}{15} - \log \frac{-\Box}{4\pi\mu^2} \right) R^{\mu\nu} \right.$$

$$\left. \left. + \frac{1}{120} R \left(-\frac{2}{n-4} - \mathbf{C} + \frac{1}{15} - \log \frac{-\Box}{4\pi\mu^2} \right) R \right] + O(R^3) \right\} . \qquad (3.87)$$

An analogous expression was obtained in the paper [223]. However, in that paper the coefficients α_k, β_k, γ_k, δ_k, and ε_k, (3.55)–(3.59), in the De Witt coefficient A_k, (3.53), were not calculated. That is why in the paper [223] it was assumed additionally that the power series in the proper time for the functions $f_i(z)$, (3.26), (3.27) and (3.62)–(3.64), converge as well as the proper time integral (1.50) at the upper limit does. Besides, in [223] only divergent and logarithmic, $\log(-\Box)$, terms in (3.87) were calculated explicitly. The complete result (3.87) together with the finite constants is obtained in present work.

The renormalized effective action for the massless field can be obtained using the ultraviolet asymptotics of the formfactors (3.43), (3.44) and (3.83)–(3.85) and substituting the renormalization parameter instead of the mass, $m^2 \to \mu^2$. This reduces just to a renormalization of the formfactors (3.38) and (3.77). Although in the massless case the finite terms in (3.87) are absorbed by the renormalization ambiguity, the finite terms in the ultraviolet asymptotics $m^2 \to 0$ of the formfactors with fixed normalization of the effective action (3.78) are essential.

Let us consider the massless field in the two-dimensional space. In this case there are no ultraviolet divergences; instead, the pole in the dimension $1/(n-2)$ in (3.86) reflects the non-local infrared divergences

$$
\begin{aligned}
\Gamma_{(1)} &= \frac{1}{4\pi(n-2)} \int \mathrm{d}^n x\, g^{1/2} \mathrm{str} \left(Q \frac{1}{(-\Box)} Q + J_\mu \frac{1}{\Box^2} J^\mu \right) \\
&+ \frac{1}{2(4\pi)} \int \mathrm{d}^2 x\, g^{1/2} \mathrm{str} \left\{ Q \left(\log \frac{-\Box}{4\pi\mu^2} + \mathbf{C} \right) \frac{1}{(-\Box)} Q \right. \\
&+ J_\mu \left(\log \frac{-\Box}{4\pi\mu^2} + \mathbf{C} - 2 \right) \frac{1}{\Box^2} J^\mu + Q \frac{1}{(-\Box)} R \\
&\left. + \frac{1}{12} \hat{1}\, R \frac{1}{(-\Box)} R + O(R^3) \right\} ,
\end{aligned} \tag{3.88}
$$

where we made use of the fact that any two-dimensional space satisfies identically the Einstein equations

$$
R_{\mu\nu} = \frac{1}{2} g_{\mu\nu} R \,. \tag{3.89}
$$

For the scalar field in the conformally invariant case (3.51), i.e., $J_\mu \equiv \nabla^\alpha \mathcal{R}_{\alpha\mu} = 0$, $Q = 0$, the effective action of the massless field in the two-dimensional space is finite up to cubic terms in background fields, (3.88), and has the form

$$
\Gamma_{(1)} = \frac{1}{24(4\pi)} \int \mathrm{d}^2 x\, g^{1/2} \left\{ R \frac{1}{(-\Box)} R + O(R^3) \right\} . \tag{3.90}
$$

On the other hand, any two-dimensional space is conformally flat. Therefore, any functional of the metric $g_{\mu\nu}$ is uniquely determined by the trace of the functional derivative. Hence, the effective action of the massless conformally invariant field in the two-dimensional space can be obtained by the integration of the conformal anomaly [223, 42, 194]. The exact answer has the form (3.90) without the third order terms $O(R^3)$, i.e., they vanish in the conformally invariant case. When the conformal invariance is absent the infrared divergences $Q\,\Box^{-1}Q$ and $J_\mu\,\Box^{-2}J^\mu$ as well as the finite terms of higher orders $O(R^3)$, (3.88), appear.

3.5 De Witt Coefficients in De Sitter Space

In the Sects. 3.3 and 3.4 we summed up the linear and quadratic terms in background fields in the asymptotic expansion of the $\Omega(s)$ in the powers of the proper time, (1.43). We showed that the corresponding series (3.26), (3.27) and (3.62)–(3.64) converge for any value of the proper time and define the entire functions (3.29), (3.30) and (3.65)–(3.68). The opposite case is the summation of those terms in the asymptotic expansion of the at coinciding points, (1.43),

$$\Omega(s|x,x) = \sum_{k\geq 0} \frac{(is)^k}{k!}[a_k]\,, \tag{3.91}$$

that do not contain the covariant derivatives of the background field. The summation of the linear and quadratic terms determines the high-energy asymptotics of the effective action ($\Box R \gg R^2$), whereas the terms without derivatives determine its low-energy asymptotics ($\Box R \ll R^2$). We limit ourselves, for simplicity, to the scalar case, i.e., we set $\mathcal{R}_{\mu\nu} = 0$.

The De Witt coefficients in the coinciding points $[a_k]$ have the following general form

$$[a_k] = \sum_{0\leq l\leq k} Q^l \underbrace{R^{\cdots\cdot}\cdots R_{\cdots\cdot}}_{k-l} + O(\nabla R)\,, \tag{3.92}$$

where the sum contains a finite number of local terms constructed from the curvature tensor and the potential term Q, and $O(\nabla R)$ denotes a finite number of local terms constructed from the curvature tensor, the potential term Q and their covariant derivatives, so that in the case of covariantly constant curvature tensor and the potential term,

$$\nabla_\mu R_{\alpha\beta\gamma\delta} = 0, \qquad \nabla_\mu Q = 0, \tag{3.93}$$

these terms vanish, $O(\nabla R) = 0$. Thus, to find the terms without covariant derivatives in the De Witt coefficients $[a_k]$, (3.92), it is sufficient to restrict oneself to the constant potential term, $Q = \text{const}$, and symmetric spaces, (3.93).

The problem of low-energy asymptotics is much more difficult than that of high-energy asymptotics. A proper treatment of this problem goes well beyond the scope of this book, since it requires the apparatus of harmonic analysis on symmetric spaces (see, for example, our recent papers [13]-[21]). Our goal in present work is to illustrate how one can employ the technique developed in Sect. 2 to get non-perturbative results. We would like to warn the reader that our result will not be exact but it will reproduce correctly the semi-classical Schwinger-De Witt expansion.

That is why we restrict ourselves for simplicity to a De Sitter space, when the background curvature is characterized by only *one* dimensional constant—the scalar curvature, (2.104),

$$R^{\mu}{}_{\alpha\nu\beta} = \frac{R}{n(n-1)}\left(\delta^{\mu}_{\nu} g_{\alpha\beta} - \delta^{\mu}_{\beta} g_{\alpha\nu}\right), \qquad R = \text{const} > 0\,. \tag{3.94}$$

Therefore, the De Witt coefficients in De Sitter space can be expressed only in terms of the scalar curvature R

$$[a_k] = \sum_{0\le l\le k} c_{k,l} R^{k-l} Q^l\,, \tag{3.95}$$

where $c_{k,l}$ are numerical coefficients. Hereafter we omit the terms $O(\nabla R)$. On the other hand, from the definition of the De Witt coefficients as the coefficients of the asymptotic expansion of the transfer function, (1.43), it is easy to get the dependence of the De Witt coefficients on the potential term Q,

$$[a_k] = \sum_{0\le l\le k} \binom{k}{l} c_{k-l} R^{k-l} Q^l\,, \tag{3.96}$$

where

$$c_k \equiv c_{k,0} = R^{-k}[a_k]\Big|_{Q=0}$$

are dimensionless De Witt coefficients computed for $Q = 0$.

The De Witt coefficients in De Sitter space can be calculated along the lines of the method developed in the Chap. 2. However, it is more convenient to find them by means of the asymptotic expansion of the Green function in the coinciding points. It was obtained in the paper [89] in any n-dimensional De Sitter space

$$G(x,x) = \frac{\mathrm{i}}{(4\pi)^{n/2}}\left(\frac{R}{n(n-1)}\right)^{\frac{n}{2}-1} \Gamma\left(1-\frac{n}{2}\right)\Phi_n(\beta)\,, \tag{3.97a}$$

$$\Phi_n(\beta) = \frac{\Gamma\left(\frac{n-1}{2}+\mathrm{i}\beta\right)\Gamma\left(\frac{n-1}{2}-\mathrm{i}\beta\right)}{\Gamma\left(\frac{1}{2}+\mathrm{i}\beta\right)\Gamma\left(\frac{1}{2}-\mathrm{i}\beta\right)}\,, \tag{3.97b}$$

where

$$\beta^2 = \frac{n(n-1)}{R}\left(m^2 - Q - \frac{n-1}{4n}R\right). \tag{3.98}$$

The asymptotic expansion of the function (3.97b) in the inverse powers of β^2 has the form

$$\Phi_n(\beta) = \sum_{k\geq 0} d_k(n)\beta^{2(\frac{n}{2}-1-k)} . \tag{3.99}$$

The coefficients $d_k(n)$ are determined by using the asymptotic expansion of the gamma-function [98] from the relation

$$\sum_{k\geq 0} d_k(n)z^k = \exp\left\{\sum_{k\geq 1} \frac{(-1)^{k+1}}{k(2k+1)} B_{2k+1}\left(\frac{n-1}{2}\right) z^k\right\} , \tag{3.100}$$

$B_k(x)$ being the Bernoulli polynomials [98], and have the form

$$d_0(n) = 1,$$

$$d_k(n) = \sum_{1\leq l\leq k} \frac{(-1)^{k+l}}{l!} \sum_{\substack{k_1,\dots,k_l\geq 1 \\ k_1+\cdots+k_l=k}} \frac{B_{2k_1+1}\left(\frac{n-1}{2}\right)}{k_1(2k_1+1)} \cdots \frac{B_{2k_l+1}\left(\frac{n-1}{2}\right)}{k_l(2k_l+1)} . \tag{3.101}$$

On the other hand, using the Schwinger–De Witt presentation of the Green function, (1.49), the proper time expansion of the transfer function, (3.91), and the equation (3.95), we obtain the asymptotic Schwinger–De Witt expansion for the Green function

$$G(x,x) = \frac{\mathrm{i}}{(4\pi)^{n/2}} \left\{\frac{R}{n(n-1)}\right\}^{\frac{n}{2}-1} \sum_{k\geq 0} \frac{\Gamma\left(k+1-\frac{n}{2}\right)}{k!} b_k(n)\beta^{2(\frac{n}{2}-1-k)} , \tag{3.102}$$

where

$$b_k(n) = \left\{\frac{R}{n(n-1)}\right\}^{-k} [a_k]\Big|_{Q=-[(n-1)/(4n)]R} \tag{3.103}$$

are the dimensionless De Witt coefficients calculated for $Q = -\frac{n-1}{4n}R$.

The total De Witt coefficients for arbitrary Q are expressed in terms of the coefficients b_k, (3.103), according to (3.95):

$$[a_k] = \sum_{0\leq l\leq k} \binom{k}{l} b_{k-l}(n) \left(\frac{R}{n(n-1)}\right)^{k-l} \left(Q + \frac{n-1}{4n}R\right)^l . \tag{3.104}$$

Comparing (3.97), (3.99) and (3.102) we obtain the coefficients $b_k(n)$

$$b_k(n) = (-1)^k k! \frac{\Gamma\left(\frac{n}{2}-k\right)}{\Gamma\left(\frac{n}{2}\right)} d_k(n) . \tag{3.105}$$

To get the coefficients $b_k(n)$ for integer values of the dimension n one has to take the limit in (3.105). Thereby it is important to take into account the dependence of the coefficients $d_k(n)$, (3.101), on the dimension n. From the definition of these coefficients, (3.97b), (3.99), one can show that in integer dimensions ($n = 2, 3, \ldots$) they vanish for $k \geq [n/2]$

$$d_k(n) = 0, \qquad \left(n = 2, 3, 4 \ldots;\ k \geq \left[\frac{n}{2}\right]\right) . \tag{3.106}$$

Let us consider first the case of odd dimension ($n = 3, 5, 7, \ldots$). In this case the gamma-function in (3.105) does not have any poles and the formula (3.105) immediately gives the coefficients $b_k(n)$. From (3.106) it follows that only first $(n-1)/2$ coefficients $b_k(n)$ do not vanish, i.e.,

$$b_k(n) = 0, \qquad \left(n = 3, 5, 7, \ldots;\ k \geq \frac{n-1}{2}\right) . \tag{3.107}$$

This is the consequence of the finiteness of the Green function (3.97) in odd dimension.

In even dimensions ($n = 2, 4, 6, \ldots$) and for $k \leq \frac{n}{2} - 1$ the expression (3.105) is also single-valued and immediately defines the coefficients $b_k(n)$. For $k \geq \frac{n}{2}$ there appear poles in the gamma-function that are suppressed by the zeros of the coefficients $d_k(n)$, (3.106). Using the definition of the coefficients $d_k(n)$, (3.100), (3.101), and the properties of the Bernoulli polynomials [98], we obtain the coefficients $b_k(n)$ in even dimension ($n = 2, 4, 6, \ldots$) for $k \geq n/2$

$$b_k(n) = \frac{(-1)^{k-(n/2)} k!}{\Gamma\left(\frac{n}{2}\right) \Gamma\left(k + 1 - \frac{n}{2}\right)} \sum_{0 \leq l \leq (n/2)-1} \frac{(-1)^l}{k-l} \left(1 - 2^{1+2l-2k}\right) B_{2k-2l} d_l(n) ,$$
$$\left(n = 2, 4, 6, \ldots;\ k \geq \frac{n}{2}\right) , \tag{3.108}$$

where $B_l = B_l(0)$ are the Bernoulli numbers and the coefficients $d_l(n)$, ($l \leq n/2 - 1$), are calculated by means of the formula (3.101).

Let us list the dimensionless De Witt coefficients $b_k(n)$, (3.103), in two- three- and four-dimensional (physical) space-time. Substituting $n = 2$ and $n = 4$ in the formulas (3.101), (3.105) and (3.108) we obtain

$$b_k(2) = (-1)^{k-1} \left(1 - 2^{1-2k}\right) B_{2k} , \qquad (k \geq 0) , \tag{3.109}$$

$$\begin{cases} b_0(4) = 1, \\ b_k(4) = (-1)^k \Big\{ (k-1) \left(1 - 2^{1-2k}\right) B_{2k} \\ \qquad - \frac{k}{4} \left(1 - 2^{3-2k}\right) B_{2k-2} \Big\} , \quad (k \geq 1) . \end{cases} \tag{3.110}$$

The expression for the coefficients $b_k(n)$ in even dimension for $k \geq n/2$, (3.108), can be written in a more convenient and compact form. Using the integral representation of the Bernoulli numbers [98]

$$B_{2k} = (-1)^{k+1} \frac{4k}{1-2^{1-2k}} \int\limits_0^\infty \frac{dt}{e^{2\pi t}+1} t^{2k-1} \tag{3.111}$$

and the definition of the coefficients $d_k(n)$, (3.97b), (3.99)–(3.101), we obtain from (3.108)

$$b_k(n) = \frac{4(-1)^{\frac{n}{2}-1} k!}{\Gamma\left(\frac{n}{2}\right)\Gamma\left(k+1-\frac{n}{2}\right)} \int\limits_0^\infty \frac{dt\, t}{e^{2\pi t}+1}$$

$$\times \frac{\Gamma\left(\frac{n-1}{2}+it\right)\Gamma\left(\frac{n-1}{2}-it\right)}{\Gamma\left(\frac{1}{2}+it\right)\Gamma\left(\frac{1}{2}-it\right)} t^{2k-n} ,$$

$$\left(n = 2, 4, 6, \ldots;\ k \geq \frac{n}{2}\right) . \tag{3.112}$$

Thus we have obtained the De Witt coefficients in De Sitter space, $[a_k]$, (3.104), where the coefficients $b_k(n)$ are given by the formulas (3.105), (3.101) and (3.108)–(3.112).

Using the obtained De Witt coefficients, (3.104), one can calculate the transfer function in the coinciding points, (3.91). Substituting (3.104) in (3.91) and summing the powers of the potential term we obtain

$$\Omega(s|x,x) = \exp\left\{is\left(Q + \frac{n-1}{4n}R\right)\right\} \omega\left(\frac{isR}{n(n-1)}\right) , \tag{3.113}$$

where

$$\omega(z) = \sum_{k\geq 0} \frac{z^k}{k!} b_k(n) . \tag{3.114}$$

Let us divide the series (3.114) in two parts

$$\omega(z) = \omega_1(z) + \omega_2(z) , \tag{3.115}$$

where

$$\omega_1(z) = \sum_{0\leq k\leq [n/2]-1} \frac{z^k}{k!} b_k(n) , \tag{3.116}$$

$$\omega_2(z) = \sum_{k\geq [n/2]} \frac{z^k}{k!} b_k(n) . \tag{3.117}$$

The first part $\omega_1(z)$, (3.116), is a polynomial. Using the expression (3.105) for coefficients b_k and the definition of the coefficients d_k, (3.97), (3.99), one can write it in the integral form

$$\omega_1(z) = 2\frac{(-z)^{\frac{n}{2}-1}}{\Gamma\left(\frac{n}{2}\right)}\int\limits_0^\infty dt\, t\exp\left(-t^2\right)\frac{\Gamma\left(\frac{n-1}{2}+\mathrm{i}\frac{t}{\sqrt{-z}}\right)\Gamma\left(\frac{n-1}{2}-\mathrm{i}\frac{t}{\sqrt{-z}}\right)}{\Gamma\left(\frac{1}{2}+\mathrm{i}\frac{t}{\sqrt{-z}}\right)\Gamma\left(\frac{1}{2}-\mathrm{i}\frac{t}{\sqrt{-z}}\right)} . \tag{3.118}$$

The second part $\omega_2(z)$, (3.117), is an asymptotic series. In odd dimension all coefficients of this series are equal to zero, (3.107). Therefore, in this case $\omega_2(z) = 0$ up to an arbitrary non-analytical function in the vicinity like $\exp(\mathrm{const}/z)$.

From the obtained expressions (3.108)–(3.112) one can get the asymptotics of the coefficients $b_k(n)$ as $k \to \infty$ in even dimensions

$$b_k(n)\Big|_{k\to\infty} = 4\frac{(-1)^{\frac{n}{2}-1}}{\Gamma\left(\frac{n}{2}\right)}\frac{k!(2k-1)!}{\Gamma\left(k+1-\frac{n}{2}\right)}(2\pi)^{-2k} , \qquad (n = 2, 4, 6, \dots) . \tag{3.119}$$

From here it is immediately seen that in this case the asymptotic series (3.117) diverges for any $z \neq 0$, i.e., its radius of convergence is equal to zero and the point $z = 0$ is a singular point of the function $\omega_2(z)$. This makes the asymptotic series in De Sitter space very different from the corresponding series of linear and quadratic terms in background fields, (3.26), (3.27), (3.62)–(3.64), which converge for any z and define entire functions, (3.29), (3.30), (3.65)–(3.68).

Based on the divergence of the asymptotic series (3.117), it is concluded in the paper [46] that the Schwinger–De Witt representation of the Green function is meaningless. However, the asymptotic divergent series (3.117) can be used to obtain quite certain idea about the function $\omega_2(z)$. To do this one has to make use of the methods for summation of asymptotic series discussed in Sect. 3.1.

Let us define the function $\omega_2(z)$ according to the formulas (3.5) and (3.3):

$$\omega_2(z) = \int\limits_0^\infty dy\, y^{\nu-1}\mathrm{e}^{-y}B(zy^\mu) , \tag{3.120}$$

where

$$B(z) = \sum_{k\geq n/2}\frac{z^k}{k!}\frac{b_k}{\Gamma(\mu k+\nu)} , \tag{3.121}$$

and μ and ν are some complex numbers, such that $\mathrm{Re}\left(\mu\frac{n}{2}+\nu\right) > 0$. The series (3.121) for the Borel function $B(z)$ converges in case $\mathrm{Re}\,\mu > 1$ for any z and in case $\mathrm{Re}\,\mu = 1$ for $|z| < \pi^2$. This can be seen easily using the formulas (3.4) and (3.119). Let us substitute the integral representation of the coefficients b_k, (3.112), in (3.121) and change the order of integration and summation. We get

$$B(z) = -4\frac{(-z)^{\frac{n}{2}}}{\Gamma\left(\frac{n}{2}\right)}\int\limits_0^\infty \frac{dt\, t}{\mathrm{e}^{2\pi t}+1}\frac{\Gamma\left(\frac{n-1}{2}+\mathrm{i}t\right)\Gamma\left(\frac{n-1}{2}-\mathrm{i}t\right)}{\Gamma\left(\frac{1}{2}+\mathrm{i}t\right)\Gamma\left(\frac{1}{2}-\mathrm{i}t\right)}H(zt^2) , \tag{3.122}$$

where

$$H(z) = \sum_{k\geq 0} \frac{z^k}{k!\Gamma\left(\mu k + \nu + \frac{n}{2}\right)} . \tag{3.123}$$

The series (3.123) converges for any z and defines an entire function. For example, for $\mu = 1$, $\nu = 1 - \frac{n}{2}$ it reduces to the Bessel function of zeroth order

$$H(z)\Big|_{\mu=1,\, \nu=1-(n/2)} = J_0\left(2\sqrt{-z}\right) . \tag{3.124}$$

One can show that the integral (3.122) converges for sure on the negative part of the real axis, $\mathrm{Re}\, z < 0$, $\mathrm{Im}\, z = 0$. Therefore, for such z it gives the analytical continuation of the series (3.121). The whole Borel function $B(z)$ on the whole complex plane can be obtained by the analytic continuation of the integral (3.122).

Let us substitute now the expression (3.122) for the Borel function in the integral (3.120) and change the order of integration over t and over y. Integrating over y and summing over k we obtain

$$\omega_2(z) = -4\frac{(-z)^{\frac{n}{2}}}{\Gamma\left(\frac{n}{2}\right)} \int\limits_0^\infty \frac{\mathrm{d}t\, t}{\mathrm{e}^{2\pi t}+1} \frac{\Gamma\left(\frac{n-1}{2}+\mathrm{i}t\right)\Gamma\left(\frac{n-1}{2}-\mathrm{i}t\right)}{\Gamma\left(\frac{1}{2}+\mathrm{i}t\right)\Gamma\left(\frac{1}{2}-\mathrm{i}t\right)} \exp(zt^2) . \tag{3.125}$$

This integral converges in the region $\mathrm{Re}\, z \leq 0$. In the other part of the complex plane the function $\omega_2(z)$ is defined by analytical continuation. In this way we obtain a single-valued analytical function in the complex plane with a cut along the positive part of the real axis from 0 to ∞. Thus the point $z = 0$ is a singular (branching) point of the function $\omega_2(z)$. It is this fact that makes the power series over z, (3.117), to diverge for any $z \neq 0$.

In the region $\mathrm{Re}\, z < 0$ one can obtain analogous expression for the total function $\omega(z)$, (3.115). Changing the integration variable $t \to t\sqrt{-z}$ in (3.118) and adding it to (3.125) we obtain

$$\omega(z) = 2\frac{(-z)^{\frac{n}{2}}}{\Gamma\left(\frac{n}{2}\right)} \int\limits_0^\infty \mathrm{d}t\, t \tanh(\pi t) \frac{\Gamma\left(\frac{n-1}{2}+\mathrm{i}t\right)\Gamma\left(\frac{n-1}{2}-\mathrm{i}t\right)}{\Gamma\left(\frac{1}{2}+\mathrm{i}t\right)\Gamma\left(\frac{1}{2}-\mathrm{i}t\right)} \exp(zt^2) . \tag{3.126}$$

Thus, we summed up the divergent asymptotic series of the terms without covariant derivatives of the background field in the transfer function, (3.113), (3.114), (3.126).

Using the obtained transfer function , (3.113), (3.118), (3.126), one can calculate the one-loop effective action, (1.50). In order to obtain the renormalized effective action $\Gamma_{(1)\mathrm{ren}}$ with the normalization condition (3.78) it is sufficient to subtract from the $\Omega(s)$, (3.113), the potentially divergent terms and integrate over the proper time. As the result we obtain in two-dimensional space

$$\Gamma_{(1)\mathrm{ren}}\Bigg|_{n=2} = \frac{1}{2(4\pi)}\int \mathrm{d}^2x\, g^{1/2}\Bigg\{\left(m^2 - Q - \frac{1}{6}R\right)\log\frac{m^2 - Q - \frac{1}{8}R}{m^2}$$

$$+Q + \frac{1}{8}R - 2R\chi_1\left(2\frac{m^2-Q}{R} - \frac{1}{4}\right)\Bigg\}, \qquad (3.127)$$

and in physical four-dimensional space-time

$$\Gamma_{(1)\mathrm{ren}}\Bigg|_{n=4} = \frac{1}{2(4\pi)^2}\int \mathrm{d}^4x\, g^{1/2}\Bigg\{-\frac{1}{2}\Bigg[m^4 - 2m^2\left(Q + \frac{1}{6}R\right)$$

$$+Q^2 + \frac{1}{3}QR + \frac{29}{1080}R^2\Bigg]\log\frac{m^2 - Q - \frac{9}{48}R}{m^2}$$

$$-\frac{1}{2}m^2\left(Q + \frac{9}{48}R\right) + \frac{5}{4}Q^2 + \frac{43}{96}QR$$

$$+\frac{41}{1024}R^2 + \frac{1}{36}R^2\chi_2\left(12\frac{m^2-Q}{R} - \frac{9}{4}\right)\Bigg\}, \qquad (3.128)$$

where

$$\chi_1(z) = \int_0^\infty \frac{\mathrm{d}t\, t}{\mathrm{e}^{2\pi t}+1}\log\left(1 - \frac{t^2}{z}\right), \qquad (3.129)$$

$$\chi_2(z) = \int_0^\infty \frac{\mathrm{d}t\, t}{\mathrm{e}^{2\pi t}+1}\left(t^2 + \frac{1}{4}\right)\log\left(1 - \frac{t^2}{z}\right). \qquad (3.130)$$

In odd dimensions the effective action is finite. Substituting (3.118), (3.113) in (1.50) and integrating over the proper time we obtain, for example, in three-dimensional space

$$\Gamma_{(1)}\Bigg|_{n=3} = \frac{1}{3(4\pi)}\int \mathrm{d}^3x\, g^{1/2}\left(m^2 - Q - \frac{1}{6}R\right)^{3/2}. \qquad (3.131)$$

Let us remind once again that our study of the low-energy effective action in this book aimed only illustrative purposes. In general case the structure of symmetric spaces is much more complicated that De Sitter space. In particular, the invariants of the curvature do not reduce to the powers of the scalar curvature but also include invariants of the Ricci and the Weyl tensors. For more accurate treatment of low-energy asymptotics see, for example, [13]–[21].

Here we only showed that by partial summation of some low-energy terms in the Schwinger–De Witt asymptotic expansion one can obtain a non-trivial expression for the effective action, (3.127)–(3.131). Although the corresponding asymptotic series diverge, the expressions (3.129) and (3.130) are defined in the whole complex plane z with the cut along the positive part of the real axis. There appears a natural arbitrariness connected with the possibility to choose the different banks of the cut.

4. Higher-Derivative Quantum Gravity

4.1 Quantization of Gauge Field Theories

Let $M = \{\varphi^i\}$ be the configuration space of a boson gauge field and $S(\varphi)$ be a classical action functional that is invariant with respect to local gauge transformations

$$\delta\varphi^i = R^i{}_\alpha(\varphi)\xi^\alpha \ , \tag{4.1}$$

forming the gauge group G. Here ξ^α are the group parameters and $R^i{}_\alpha(\varphi)$ are the local generators of the gauge transformations that form a closed Lie algebra

$$[\hat{\mathbf{R}}_\alpha, \hat{\mathbf{R}}_\beta] = C^\gamma{}_{\alpha\beta}\hat{\mathbf{R}}_\gamma \ , \tag{4.2}$$

where

$$\hat{\mathbf{R}}_\alpha \equiv R^i{}_\alpha(\varphi)\frac{\delta}{\delta\varphi^i} \ , \tag{4.3}$$

and $C^\gamma{}_{\alpha\beta}$ are the structure constants of the gauge group satisfying the Jacobi identity

$$C^\alpha{}_{\mu[\beta}C^\mu{}_{\gamma\delta]} = 0 \ . \tag{4.4}$$

The classical equations of motion determined by the action functional $S(\varphi)$ have the form

$$\varepsilon_i(\varphi) = 0 \ , \tag{4.5}$$

where $\varepsilon_i \equiv S_{,i}$ is the "extremal" of the action. The equation (4.5) defines the "mass shell" in the quantum perturbation theory.

The equations (4.5) are not independent. The gauge invariance of the action leads to the first type constraints between the dynamical variables that are expressed through the Nöther identities

$$\hat{\mathbf{R}}_\alpha S = R^i{}_\alpha\varepsilon_i = 0 \ . \tag{4.6}$$

The physical dynamical variables are the group orbits (the classes of gauge equivalent field configurations), and the physical configuration space is the space of orbits $\mathcal{M} = M/G$. To have coordinates on the orbit space one has to put some supplementary gauge conditions (a set of constraints) that isolate in the space M a subspace $\mathcal{M}'$, which intersects each orbit only in one point. Each orbit is represented then by the point in which it intersects the given subspace $\mathcal{M}'$. If one reparametrizes the initial configuration space by the new variables, $M = \{I^A, \chi_\mu\}$, where I^A are the physical gauge-invariant variables enumerating the orbits, $\mathcal{M} = \{I^A\}$, and χ_μ are group variables that enumerate the points on each orbit, $G = \{\chi_\mu\}$, then the gauge conditions can be written in the form

$$\chi_\mu(\varphi) = \theta_\mu \ ,$$

where θ_μ are some constants, i.e., $\theta_{\mu,i} = 0$. Thus we obtain the coordinates on the physical configuration space $\mathcal{M}' = \{I^A, \theta_\mu\}$.

The group variables χ_μ transform under the action of the gauge group analogously to (4.1)

$$\delta\chi_\mu = Q_{\mu\alpha}(\varphi)\xi^\alpha \ , \tag{4.7}$$

where

$$Q_{\mu\alpha}(\varphi) = \chi_{\mu,i}(\varphi)R^i{}_\alpha(\varphi) \ , \qquad (\det Q \neq 0) \ , \tag{4.8}$$

are the generators of gauge transformations of group variables. They form a representation of the Lie algebra of the gauge group

$$[\hat{\mathbf{Q}}_\alpha, \hat{\mathbf{Q}}_\beta] = C^\gamma{}_{\alpha\beta}\hat{\mathbf{Q}}_\gamma \ , \tag{4.9}$$

where

$$\hat{\mathbf{Q}}_\alpha \equiv Q_{\mu\alpha}\frac{\delta}{\delta\chi_\mu} \ .$$

All physical gauge-invariant quantities (in particular, the action $S(\varphi)$) are expressed only in terms of the gauge-invariant variables I^A and do not depend on the group variables χ_μ

$$S(\varphi) = \bar{S}(I^A(\varphi)) \ . \tag{4.10}$$

The action $\bar{S}(I^A)$ is an usual non-gauge-invariant action. Therefore one can quantize the theory in the variables I^A, and then go back to the initial field variables φ^i. The functional integral for the standard effective action takes the form [223]

$$\exp\left\{\frac{\mathrm{i}}{\hbar}\Gamma(\Phi)\right\} = \int \mathrm{d}\varphi\, \mathcal{M}(\varphi)\delta(\chi_\mu(\varphi) - \theta_\mu) \det Q(\varphi)$$

$$\times \exp\left\{\frac{\mathrm{i}}{\hbar}\left[S(\varphi) - (\varphi^i - \Phi^i)\Gamma_{,i}(\Phi)\right]\right\} \ , \tag{4.11}$$

where $\mathcal{M}(\varphi)$ is a local measure. The measure $\mathcal{M}(\varphi)$ is gauge-invariant up to the terms $R^i{}_{\alpha,i}$ and $C^\mu{}_{\alpha\mu}$ that are proportional to derivatives of the delta-function at coinciding points $\delta(0)$ and that should vanish by the regularization. The exact form of the measure $\mathcal{M}(\varphi)$ must be determined by the canonical quantization of the theory. In most practically important cases $\mathcal{M}(\varphi) = 1 + \delta(0)(\ldots)$ [223]. Therefore, below we will simply omit the measure $\mathcal{M}(\varphi)$.

Since the effective action (4.11) must not depend on the arbitrary constants θ_μ, one can integrate over them with a Gaussian weight. As result we get

$$\exp\left\{\frac{\mathrm{i}}{\hbar}\Gamma(\Phi)\right\} = \int d\varphi \, \det Q(\varphi)(\det H)^{1/2}$$

$$\times \exp\left\{\frac{\mathrm{i}}{\hbar}\left[S(\varphi) - \frac{1}{2}\chi_\mu(\varphi)H^{\mu\nu}\chi_\nu(\varphi) - (\varphi^i - \Phi^i)\Gamma_{,i}(\Phi)\right]\right\} ,$$

$$(4.12)$$

where $H^{\mu\nu}$ is a non-degenerate operator ($\det H \neq 0$), that does not depend on quantum field ($\delta H^{\mu\nu}/\delta\varphi^i = 0$). The determinants of the operators Q and H in (4.12) can be also represented as result of integration over the anti-commuting variables, so called Faddeev–Popov [101, 78] and Nielsen–Kallosh [182, 162] "ghosts".

Using the equation (4.12) we find the one-loop effective action

$$\Gamma = S + \hbar\Gamma_{(1)} + O(\hbar^2) , \tag{4.13}$$

$$\Gamma_{(1)} = -\frac{1}{2\mathrm{i}}\log\frac{\det \Delta}{\det H (\det F)^2} , \tag{4.14}$$

where

$$\Delta_{ik} = -S_{,ik} + \chi_{\mu i}H^{\mu\nu}\chi_{\nu k} , \qquad \chi_{\mu i} \equiv \chi_{\mu,i}(\varphi)\Big|_{\varphi=\Phi} , \tag{4.15}$$

$$F = Q(\varphi)\Big|_{\varphi=\Phi} . \tag{4.16}$$

On the mass shell, $\Gamma_{,i} = 0$, the standard effective action, (4.12), and therefore the S-matrix too, does not depend neither on the gauge (i.e., on the choice of arbitrary functions χ_μ and $H^{\mu\nu}$) nor on the parametrization of the quantum field [84, 51, 223]. However, off mass shell the background field as well as the Green functions and the effective action crucially depend both on the gauge fixing and on the parametrization of the quantum field. Moreover, the effective action is not, generally speaking, a gauge-invariant functional of the background field because the usual procedure of the gauge fixing of the quantum field automatically fixes the gauge of the background field.

The gauge invariance of the effective action off mass shell can be preserved by using the De Witt's background field gauge. In this gauge the functions $\chi_\mu(\varphi)$ and the matrix $H^{\mu\nu}$ depend parametrically on the background field [78, 84]

$$\chi_\mu(\varphi) = \chi_\mu(\varphi, \Phi) , \qquad \chi_\mu(\Phi, \Phi) = 0 , \qquad H^{\mu\nu} = H^{\mu\nu}(\Phi) ,$$

and are covariant with respect to simultaneous gauge transformations of the quantum φ and background Φ fields, i.e., they form the adjoint representation of the gauge group

$$\hat{\mathbf{R}}_\alpha(\varphi)\chi_\mu(\varphi,\Phi) + \hat{\mathbf{R}}_\alpha(\Phi)\chi_\mu(\varphi,\Phi) + C^\nu{}_{\mu\alpha}\chi_\nu(\varphi,\Phi) = 0 ,$$

$$\hat{\mathbf{R}}_\alpha(\Phi)H^{\mu\nu}(\Phi) - C^\mu{}_{\beta\alpha}H^{\beta\nu}(\Phi) - C^\nu{}_{\beta\alpha}H^{\mu\beta}(\Phi) = 0 . \qquad (4.17)$$

In the background field gauge method one fixes the gauge of the quantum field but not the gauge of the background field. In this case the standard effective action (4.12) will be gauge-invariant functional of the background field provided the generators of gauge transformations $R^i{}_\alpha(\varphi)$ are linear in the fields, i.e., $R^i{}_{\alpha,kn}(\varphi) = 0$ [84].

However, the off-shell effective action still depends parametrically on the choice of the gauge (i.e., on the functions χ_μ, and $H^{\mu,\nu}$), as well as on the parametrization of the quantum field φ. As a matter of fact this is the same problem, since the change of the gauge is in essence a reparametrization of the physical space of orbits $\mathcal{M}'$ [223]. Consequently we do not have any unique effective action off mass shell.

The off-shell effective action can give much more information about quantum processes than just the S-matrix. It should give the effective equations for the background field with all quantum corrections, $\Gamma_{,i}(\Phi) = 0$.

A possible solution of the problem of constructing the off-shell effective action was proposed by Vilkovisky [223, 224] both for usual and the gauge field theories. It was further improved by De Witt [85, 86], but we will not discuss this improvement here. For extensive updated bibliography see, for example, [53].

The condition of the reparametrization invariance of the functional $S(\varphi)$ means that $S(\varphi)$ should be a scalar on the configuration space M which is treated as a manifold with coordinates φ^i. The non-covariance of the effective action $\Gamma(\varphi)$ can be traced to the source term $(\varphi^i - \Phi^i)\Gamma_{,i}(\Phi)$ in (4.12) as the difference of coordinates $(\varphi^i - \Phi^i)$ is not a geometric object. One can achieve covariance if one replaces this difference by a two-point function, $\sigma^i(\varphi,\Phi)$, that transforms like a vector with respect to transformations of the background field Φ and like a scalar with respect to transformations of the quantum field. This quantity can be constructed by introducing a symmetric gauge-invariant connection $\Gamma^i{}_{mn}(\varphi)$ on M and by identifying $\sigma^i(\varphi,\Phi)$ with the tangent vector at the point Φ to the geodesic connecting the points φ and Φ. More precisely, $\sigma^i(\varphi,\Phi)$ is defined to be the solution of the equation

$$\sigma^k\nabla_k\sigma^i = \sigma^i , \qquad (4.18)$$

where

$$\nabla_k\sigma^i = \frac{\delta}{\delta\Phi^k}\sigma^i + \Gamma^i{}_{km}(\Phi)\sigma^m , \qquad (4.19)$$

with the boundary condition

$$\sigma^i\Big|_{\varphi=\Phi} = 0 \,. \tag{4.20}$$

Thus we obtain a unique effective action $\tilde{\Gamma}(\Phi)$ according to Vilkovisky

$$\exp\left\{\frac{\mathrm{i}}{\hbar}\tilde{\Gamma}(\Phi)\right\} = \int \mathrm{d}\varphi\,(\det H(\Phi))^{1/2} \det Q(\varphi, \Phi)$$

$$\times \exp\left\{\frac{\mathrm{i}}{\hbar}\left[S(\varphi) - \frac{1}{2}\chi_\mu(\varphi,\Phi)H^{\mu\nu}(\Phi)\chi_\nu(\varphi,\Phi) + \sigma^i(\varphi,\Phi)\tilde{\Gamma}_{,i}(\Phi)\right]\right\} \,, \tag{4.21}$$

Since the integrand in (4.21) is a scalar on M, the equation (4.21) defines a reparametrization-invariant effective action. Let us note, that such a modification corresponds to the definition of the background field Φ in a manifestly reparametrization invariant way,

$$< \sigma^i(\varphi, \Phi) >= 0 \,, \tag{4.22}$$

instead of the usual definition $\Phi^i =< \varphi^i >$.

The construction of the perturbation theory is performed by the change of the variables in the functional integral, $\varphi^i \to \sigma^i(\varphi, \Phi)$, and by expanding all the functionals in the covariant Taylor series

$$S(\varphi) = \sum_{k\geq 0} \frac{(-1)^k}{k!}\sigma^{i_1}\cdots\sigma^{i_k}\left[\nabla_{(i_1}\cdots\nabla_{i_k)}S(\varphi)\right]\Big|_{\varphi=\Phi} \,. \tag{4.23}$$

The diagrammatic technique for the usual effective action results from the substitution of the covariant functional derivatives ∇_i instead of the usual ones in the expression for the standard effective action (up to the terms $\sim \delta(0)$ that are caused by the Jacobian of the change of variables).

In particular, in the one-loop approximation, (4.13)–(4.16),

$$\tilde{\Gamma}_{(1)} = -\frac{1}{2\mathrm{i}} \log \frac{\det \tilde{\Delta}}{\det H (\det F)^2} \,, \tag{4.24}$$

where

$$\tilde{\Delta}_{ik} = -\nabla_i\nabla_k S + \chi_{\mu i}H^{\mu\nu}\chi_{\nu k} \,. \tag{4.25}$$

To construct the connection on the physical configuration space $\mathcal{M}'$ let us introduce, first of all, a non-degenerate gauge-invariant metric $E_{ik}(\varphi)$ in the initial configuration space M that satisfies the Killing equations

$$\mathcal{D}_m R_{n\alpha} + \mathcal{D}_n R_{m\alpha} = 0, \tag{4.26}$$

where $R_{m\alpha} = E_{mk}R^k{}_\alpha$ and $\mathcal{D}_m$ means the covariant derivative with Christoffel connection of the metric E_{ik}

$$\left\{ {}^{i}_{jk} \right\} = \frac{1}{2} E^{-1im} (E_{mj,k} + E_{mk,j} - E_{jk,m}) \, . \tag{4.27}$$

The metric $E_{ik}(\varphi)$ must ensure the non-degeneracy of the matrix

$$N_{\mu\nu} = R^i{}_\mu E_{ik} R^k{}_\nu \, , \qquad (\det N \neq 0) \, . \tag{4.28}$$

This enables one to define the De Witt projector [78, 83]

$$\Pi_\perp{}^m_i = \delta^m_i - R^m_\alpha B^\alpha_i \, , \qquad \Pi^2_\perp = \Pi_\perp \, , \tag{4.29}$$

where

$$B^\mu{}_n = N^{-1\,\mu\alpha} R^i{}_\alpha E_{in} \, . \tag{4.30}$$

The projector $\Pi_\perp$ projects every orbit to one point,

$$\Pi_\perp{}^m_i R^i{}_\alpha = 0 \, , \tag{4.31}$$

and is orthogonal to the generators $R^k{}_\alpha$ in the metric E_{ik}

$$R^k{}_\beta E_{km} \Pi_\perp{}^m_i = 0 \, . \tag{4.32}$$

Therefore, the subspace $\Pi_\perp M$ can be identified with the space of orbits.

The natural physical conditions lead to the following form of the connection on the configuration space [223, 224]

$$\Gamma^i{}_{mn} = \{ {}^i{}_{mn} \} + T^i{}_{mn} \, , \tag{4.33}$$

where

$$T^i{}_{mn} = -2 B^\mu_{(m} \mathcal{D}_{n)} R^i{}_\mu + B^\alpha{}_{(m} B^\beta{}_{n)} R^k{}_\beta \mathcal{D}_k R^i{}_\alpha \, . \tag{4.34}$$

The main property of the connection (4.33) is

$$\nabla_n R^i{}_\alpha \propto R^i{}_\alpha \, . \tag{4.35}$$

It means that the transformation of the quantity σ^i under the gauge transformations of both the quantum and background field is proportional to $R^i{}_\alpha(\Phi)$, and, therefore, ensures the gauge invariance of the term $\sigma^i \tilde{\Gamma}_{,i}$ in (4.21). This leads then to the fact that the Vilkovisky's effective action $\tilde{\Gamma}(\Phi)$ is reparametrization invariant and does not depend on gauge fixing off mass shell.

On the other hand, it is obvious that on mass shell the Vilkovisky's effective action coincides with the standard one,

$$\tilde{\Gamma}(\Phi)\Big|_{\text{on-shell}} = \Gamma(\Phi)\Big|_{\text{on-shell}} \, ,$$

and leads to the standard S-matrix [223].

To calculate the Vilkovisky's effective action one can choose any gauge. The result will be the same. It is convenient to use the orthogonal gauge, i.e., simply put the non-physical group variables equal to zero,

$$\chi_\mu(\varphi, \Phi) = R^i{}_\mu(\Phi) E_{ik}(\Phi) \sigma^k(\varphi, \Phi) = 0 \,, \tag{4.36}$$

i.e., $\sigma^i = \sigma^i_\perp$, where $\sigma^i_\perp = \Pi_\perp{}^i{}_n \sigma^n$. The non-metric part of the connection $T^i{}_{mn}$ satisfies the equation

$$\Pi_\perp{}^m{}_k T^i{}_{mn} \Pi_\perp{}^n{}_j = 0 \,. \tag{4.37}$$

Using this equation one can show that it does not contribute to the quantity $\sigma^i_\perp$. Therefore in the orthogonal gauge (4.36) the quantity $T^i{}_{mn}$ in the connection (4.33) can be omitted. As a result we obtain for the Vilkovisky's effective action the equation

$$\exp\left\{\frac{\mathrm{i}}{\hbar}\tilde{\Gamma}(\Phi)\right\} = \int \mathrm{d}\varphi\, \delta\left(R_{i\mu}(\Phi)\sigma^i(\varphi, \Phi)\right) \det \bar{Q}(\varphi, \Phi)$$

$$\times \exp\left\{\frac{\mathrm{i}}{\hbar}\left[S(\varphi) + \sigma^i(\varphi, \Phi)\tilde{\Gamma}_{,i}(\Phi)\right]\right\} \,, \tag{4.38}$$

where

$$\bar{Q}_{\mu\nu}(\varphi, \Phi) = R_{i\mu}(\Phi) R^i{}_\nu(\varphi) \,.$$

The change of variables $\varphi \to \sigma^i(\varphi, \Phi)$ and covariant expansions of all functionals of the form

$$S(\varphi) = \sum_{k \geq 0} \frac{(-1)^k}{k!} \sigma^{i_1} \cdots \sigma^{i_k} \left[\mathcal{D}_{(i_1} \cdots \mathcal{D}_{i_k)} S(\varphi)\right]\Big|_{\varphi = \Phi} \tag{4.39}$$

lead (up to terms $\sim \delta(0)$) to the standard perturbation theory with simple replacement of usual functional derivatives by the covariant ones with the Christoffel connection, $\mathcal{D}_i$. In particular, in the one-loop approximation we have

$$\tilde{\Gamma}_{(1)} = -\frac{1}{2\mathrm{i}} \log \frac{\det \tilde{\Delta}_\perp}{(\det N)^2} \,, \tag{4.40}$$

where

$$\tilde{\Delta}_\perp = \Pi_\perp \tilde{\Delta} \Pi_\perp \,,$$

$$\tilde{\Delta}_{ik} = -\mathcal{D}_i \mathcal{D}_k S = -S_{,ik} + \left\{{}^j{}_{ik}\right\} S_{,j} \,. \tag{4.41}$$

4.2 One-Loop Divergences in Minimal Gauge

The theory of gravity with higher derivatives as well as the Einstein gravity and any other metric theory of gravity is a non-Abelian gauge theory with the group of diffeomorphisms G of the space-time as the gauge group. The complete configuration space M is the space of all pseudo-Riemannian metrics on the space-time, and the physical configuration space, i.e., the space of orbits $\mathcal{M}$, is the space of geometries on the space-time.

We will parametrize the gravitational field by the metric tensor of the space-time

$$\varphi^i = g_{\mu\nu}(x) \,, \qquad i \equiv (\mu\nu, x) \,. \tag{4.42}$$

The parameters of the gauge transformation are the components of the vector of the infinitesimal diffeomorphism, (a general coordinate transformation)

$$x^\mu \to x^\mu - \xi^\mu(x) \,,$$
$$\xi^\mu = \xi^\mu(x) \,, \qquad \mu \equiv (\mu, x) \,. \tag{4.43}$$

The local generators of the gauge transformations in the parametrization (4.42) are linear in the fields, i.e., ${R^i}_{\alpha,kn} = 0$, and have the form

$${R^i}_\alpha = 2\nabla_{(\mu} g_{\nu)\alpha} \delta(x,y) \,, \qquad i \equiv (\mu\nu, x) \,, \qquad \alpha \equiv (\alpha, y) \,. \tag{4.44}$$

Here and below, when writing the kernels of the differential operators, all the derivatives act on the first argument of the delta-function.

The local metric tensor of the configuration space, that satisfies the Killing equations (4.26), has the form

$$E_{ik} = g^{1/2} E^{\mu\nu,\alpha\beta} \delta(x,y) \,, \qquad i \equiv (\mu\nu, x) \,, \qquad k \equiv (\alpha\beta, y) \,,$$
$$E^{\mu\nu,\alpha\beta} = g^{\mu(\alpha} g^{\beta)\nu} - \frac{1}{4}(1+\kappa) g^{\mu\nu} g^{\alpha\beta} \,, \tag{4.45}$$

where $\kappa \neq 0$ is a numerical parameter. In the four-dimensional pseudo-Euclidean space-time it is equal to the determinant of the 10×10 matrix:

$$\kappa = \det\left(g^{1/2} E^{\mu\nu,\alpha\beta}\right) \,. \tag{4.46}$$

The Christoffel symbols of the metric E_{ik} have the form

$$\left\{ {i \atop jk} \right\} = \left\{ {\mu\nu,\alpha\beta \atop \lambda\rho} \right\} \delta(x,y)\delta(x,z) \,,$$
$$i \equiv (\lambda\rho, x) \,, \qquad j \equiv (\mu\nu, y) \,, \qquad k \equiv (\alpha\beta, z) \,,$$
$$\left\{ {\mu\nu,\alpha\beta \atop \lambda\rho} \right\} = -\delta^{(\mu}_{(\lambda} g^{\nu)(\alpha} \delta^{\beta)}_{\rho)} + \frac{1}{4}\left(g^{\mu\nu} \delta^{\alpha\beta}_{\lambda\rho} + g^{\alpha\beta} \delta^{\mu\nu}_{\lambda\rho} + \kappa^{-1} E^{\mu\nu,\alpha\beta} g_{\lambda\rho} \right) \,,$$
$$\delta^{\mu\nu}_{\alpha\beta} \equiv \delta^\mu_{(\alpha} \delta^\nu_{\beta)} \,. \tag{4.47}$$

The operator $N_{\mu\nu}(\varphi)$, (4.28), is a second order differential operator of the form

$$N_{\mu\nu}(\varphi) = 2g^{1/2} \left\{ -g_{\mu\nu} \Box + \frac{1}{2}(\kappa - 1)\nabla_\mu \nabla_\nu - R_{\mu\nu} \right\} \delta(x,y) \,. \tag{4.48}$$

The condition that the operator $N_{\mu\nu}$, (4.48), should be non-degenerate in the perturbation theory on the flat background, $\det N\big|_{R=0} \neq 0$, imposes a constraint on the parameter of the metric, $\kappa \neq 3$.

The classical action of gravity theory with quadratic terms in the curvature has the following general form

$$S(\varphi) = -\int \mathrm{d}^4 x g^{1/2} \left\{ \epsilon R^* R^* + \frac{1}{2f^2} C^2 - \frac{1}{6\nu^2} R^2 - \frac{1}{k^2} R + 2\frac{\lambda}{k^4} \right\}, \quad (4.49)$$

where $R^* R^* \equiv \frac{1}{4} \varepsilon^{\mu\nu\alpha\beta} \varepsilon_{\lambda\rho\gamma\delta} R^{\lambda\rho}{}_{\mu\nu} R^{\gamma\delta}{}_{\alpha\beta}$ is the topological term, $\varepsilon_{\mu\nu\alpha\beta}$ is the anti-symmetric Levi–Civita tensor, $C^2 \equiv C^{\mu\nu\alpha\beta} C_{\mu\nu\alpha\beta}$ is the square of the Weyl tensor, ϵ is the topological coupling constant, f^2 — the Weyl coupling constant, ν^2 — the conformal coupling constant, k^2 — the Einstein coupling constant, and $\lambda = \Lambda k^2$ is the dimensionless cosmological constant. Here and below we omit the surface terms $\sim \Box R$ that do not contribute neither to the equations of motion nor to the S-matrix.

The extremal of the classical action, (4.5), and the Nöther identities, (4.6), have the form

$$\varepsilon_i(\varphi) = \varepsilon^{\mu\nu} \equiv \frac{\delta S}{\delta g_{\mu\nu}} = -g^{1/2} \left\{ \frac{1}{k^2} \left(R^{\mu\nu} - \frac{1}{2} g^{\mu\nu} R + \Lambda g^{\mu\nu} \right) \right.$$

$$+ \frac{1}{f^2} \left[\frac{2}{3} (1+\omega) R (R^{\mu\nu} - \frac{1}{4} g^{\mu\nu} R) + \frac{1}{2} g^{\mu\nu} R_{\alpha\beta} R^{\alpha\beta} - 2 R^{\alpha\beta} R^{\mu}{}_{\alpha}{}^{\nu}{}_{\beta} \right.$$

$$\left. \left. + \frac{1}{3} (1 - 2\omega) \nabla^\mu \nabla^\nu R - \Box R^{\mu\nu} + \frac{1}{6} (1 + 4\omega) g^{\mu\nu} \Box R \right] \right\}, \quad (4.50)$$

$$\nabla_\mu \varepsilon^{\mu\nu} = 0, \quad (4.51)$$

where $\omega \equiv f^2/(2\nu^2)$.

Let us calculate the one-loop divergences of the standard effective action (4.14). From (4.14) we have up to the terms $\sim \delta(0)$

$$\Gamma^{\mathrm{div}}_{(1)} = -\frac{1}{2\mathrm{i}} \left(\log \det \Delta\big|^{\mathrm{div}} - \log \det H\big|^{\mathrm{div}} - 2 \log \det F\big|^{\mathrm{div}} \right). \quad (4.52)$$

The second variation of the action (4.49) has the form

$$-S_{,ik} = \left\{ \hat{F}_{(0)\lambda\rho\sigma\tau} \nabla^\lambda \nabla^\rho \nabla^\sigma \nabla^\tau + \nabla^\rho \hat{F}_{(2)\rho\sigma} \nabla^\sigma \right.$$

$$\left. + \hat{F}_{(3)\rho} \nabla^\rho + \nabla^\rho \hat{F}_{(3)\rho} + \hat{F}_{(4)} \right\} g^{1/2} \delta(x,y), \quad (4.53)$$

where $\hat{F}_{(0)\lambda\rho\sigma\tau}$, $\hat{F}_{(2)\rho\sigma}$, $\hat{F}_{(3)\rho}$ and $\hat{F}_{(4)}$ are the tensor 10×10 matrices ($\hat{F}_{(0)\lambda\rho\sigma\tau} = F^{\mu\nu,\alpha\beta}_{(0)\lambda\rho\sigma\tau}$, etc.). They satisfy the following symmetry relations

$$\hat{F}^{T}_{(2)\rho\sigma} = \hat{F}_{(2)\rho\sigma} \,, \qquad \hat{F}^{T}_{(3)\rho} = -\hat{F}_{(3)\rho} \,, \qquad \hat{F}^{T}_{(4)} = \hat{F}_{(4)} \,,$$
$$\hat{F}_{(2)\rho\sigma} = \hat{F}_{(2)\sigma\rho} \,, \tag{4.54}$$

with the symbol 'T' meaning the transposition, and have the form

$$F^{\mu\nu,\alpha\beta}_{(0)\lambda\rho\sigma\tau} = \frac{1}{2f^2}\left\{ g_{\lambda\rho}g_{\sigma\tau}\left(g^{\mu(\alpha}g^{\beta)\nu} - \frac{1+4\omega}{3}g^{\mu\nu}g^{\alpha\beta}\right)\right.$$
$$+\frac{1+4\omega}{3}\left(g^{\mu\nu}\delta^{\alpha\beta}_{\sigma\tau}g_{\lambda\rho} + g^{\alpha\beta}\delta^{\mu\nu}_{\lambda\rho}g_{\sigma\tau}\right)$$
$$\left.+\frac{2}{3}(1-2\omega)\delta^{\mu\nu}_{\lambda\rho}\delta^{\alpha\beta}_{\sigma\tau} - 2\delta^{(\nu}_{\lambda}g^{\mu)(\alpha}\delta^{\beta)}_{\tau}g_{\sigma\rho}\right\},$$

$$F^{\mu\nu,\alpha\beta}_{(2)\rho\sigma} = \frac{1}{2f^2}\left\{ 2R_{\rho\sigma}\left(g^{\mu(\alpha}g^{\beta)\nu} - \frac{1}{2}g^{\mu\nu}g^{\alpha\beta}\right) + 4g_{\rho\sigma}R^{(\mu|\alpha|\nu)\beta}\right.$$
$$+2\delta^{(\mu}_{(\sigma}R^{\nu)(\alpha}\delta^{\beta)}_{\rho)} + 2\left(g^{\alpha\beta}\delta^{(\mu}_{(\rho}R^{\nu)}_{\sigma)} + g^{\mu\nu}\delta^{(\alpha}_{(\rho}R^{\beta)}_{\sigma)}\right)$$
$$-4\left(R^{(\mu}_{(\rho}g^{\nu)(\alpha}\delta^{\beta)}_{\sigma)} + R^{(\alpha}_{(\rho}g^{\beta)(\mu}\delta^{\nu)}_{\sigma)}\right)$$
$$+\frac{4}{3}(1+\omega)\left[R^{\mu\nu}\left(\delta^{\alpha\beta}_{\rho\sigma} - g^{\alpha\beta}g_{\rho\sigma}\right) + R^{\alpha\beta}\left(\delta^{\mu\nu}_{\rho\sigma} - g^{\mu\nu}g_{\rho\sigma}\right)\right]$$
$$+\left(m_2^2 + \frac{2}{3}(1+\omega)R\right)\left(-g_{\rho\sigma}g^{\alpha(\mu}g^{\nu)\beta} + g_{\rho\sigma}g^{\alpha\beta}g^{\mu\nu}\right.$$
$$\left.\left.-g^{\mu\nu}\delta^{\alpha\beta}_{\rho\sigma} - g^{\alpha\beta}\delta^{\mu\nu}_{\rho\sigma} + 2\delta^{(\beta}_{(\sigma}g^{\alpha)(\mu}\delta^{\nu)}_{\rho)}\right)\right\},$$

$$F^{\mu\nu,\alpha\beta}_{(3)\rho} = \frac{1}{2f^2}\left\{\frac{1}{2}\left(g^{\alpha\beta}\nabla^{(\mu}R^{\nu)}_{\rho} - g^{\mu\nu}\nabla^{(\alpha}R^{\beta)}_{\rho}\right)\right.$$
$$+\frac{1}{3}(1-2\omega)\left(g^{\mu\nu}\nabla_{\rho}R^{\alpha\beta} - g^{\alpha\beta}\nabla_{\rho}R^{\mu\nu}\right) + g^{(\beta(\mu}\nabla^{\nu)}R^{\alpha)}_{\rho}$$
$$-g^{(\mu(\beta}\nabla^{\alpha)}R^{\nu)}_{\rho} + \frac{1}{12}(1+4\omega)\left(g^{\alpha\beta}\delta^{(\mu}_{\rho}\nabla^{\nu)}R - g^{\mu\nu}\delta^{(\alpha}_{\rho}\nabla^{\beta)}R\right)$$

$$+\frac{1}{6}(1-2\omega)\left(\delta_\rho^{(\alpha}g^{\beta)(\mu}\nabla^{\nu)}R-\delta_\rho^{(\mu}g^{\nu)(\alpha}\nabla^{\beta)}R\right)$$

$$+\frac{2}{3}(2-\omega)\left(\delta_\rho^{(\beta}\nabla^{\alpha)}R^{\mu\nu}-\delta_\rho^{(\mu}\nabla^{\nu)}R^{\alpha\beta}\right)$$

$$+\frac{3}{2}\left(\nabla^{(\beta}R^{\alpha)(\mu}\delta_\rho^{\nu)}-\nabla^{(\mu}R^{\nu)(\alpha}\delta_\rho^{\beta)}\right)\Bigg\},$$

$$F_{(4)}^{\mu\nu,\alpha\beta}=\frac{1}{2f^2}\Bigg\{-\frac{1}{2}\left(g^{\alpha\beta}R^\mu_\rho R^{\nu\rho}+g^{\mu\nu}R^\alpha_\rho R^{\beta\rho}\right)+4R^{(\mu\ \ \nu)}_{\ \ \rho\ \ \sigma}R^{\rho\alpha\sigma\beta}$$

$$-\left(g^{\mu(\alpha}g^{\beta)\nu}-\frac{1}{2}g^{\mu\nu}g^{\alpha\beta}\right)\Big[R_{\rho\sigma}R^{\rho\sigma}-\frac{1}{3}(1+\omega)R^2$$

$$+\frac{1}{3}(1+4\omega)\Box R-m_2^2(R-2\Lambda)\Big]$$

$$-\frac{3}{2}R^{\rho\sigma}\left(g^{\alpha\beta}R^{\mu\ \nu}_{\ \rho\ \sigma}+g^{\mu\nu}R^{\alpha\ \beta}_{\ \rho\ \sigma}\right)+R^{\alpha(\mu}R^{\nu)\beta}$$

$$+\left(m_2^2+\frac{2}{3}(1+\omega)R\right)\Big(R^{\mu\nu}g^{\alpha\beta}+R^{\alpha\beta}g^{\mu\nu}$$

$$-3g^{(\alpha(\mu}R^{\nu)\beta)}-R^{(\mu|\alpha|\nu)\beta}\Big)$$

$$+6g^{(\beta(\nu}R^{\mu)\ \ \alpha)}_{\ \ \rho\ \ \sigma}R^{\rho\sigma}+\frac{1}{2}\left(R^{(\alpha\ \ \beta)(\mu}_{\ \ \rho}R^{\nu)\rho}+R^{(\mu\ \ \nu)(\alpha}_{\ \ \rho}R^{\beta)\rho}\right)$$

$$-\frac{4}{3}(1+\omega)R^{\mu\nu}R^{\alpha\beta}-\omega\left(g^{\alpha\beta}\nabla^\mu\nabla^\nu R+g^{\mu\nu}\nabla^\alpha\nabla^\beta R\right)$$

$$+2\Box R^{(\mu|\alpha|\nu)\beta}-\frac{4}{3}(1-2\omega)g^{(\beta(\nu}\nabla^{\mu)}\nabla^{\alpha)}R$$

$$+2\Box R^{(\mu(\alpha}g^{\beta)\nu)}+\frac{2}{3}(1+\omega)\Big(\nabla^{(\alpha}\nabla^{\beta)}R^{\mu\nu}$$

$$+\nabla^{(\mu}\nabla^{\nu)}R^{\alpha\beta}-g^{\alpha\beta}\Box R^{\mu\nu}-g^{\mu\nu}\Box R^{\alpha\beta}\Big)\Bigg\}, \tag{4.55}$$

where $m_2^2=f^2/k^2$.

Next let us choose the most general linear covariant De Witt gauge condition [83]

$$\chi_\mu(\varphi, \Phi) = R^k{}_\mu(\Phi) E_{ki}(\Phi) h^i , \qquad 4.56$$

where $h^i = \varphi^i - \Phi^i$. In usual notation this condition reads

$$\chi_\mu = -2 g^{1/2} \nabla_\nu \left\{ h^\nu_\mu - \frac{1}{4}(1+\kappa)\delta^\nu_\mu h \right\} ,$$

where raising and lowering the indices as well as the covariant derivative are defined by means of the background metric $g_{\mu\nu}$ and $h \equiv h^\mu_\mu$. The ghost operator F, (4.8), (4.16), in the gauge (4.56) is equal to the operator N, (4.48),

$$F_{\mu\nu} = N_{\mu\nu} = 2 g^{1/2} \left\{ -g_{\mu\nu} \Box + \frac{1}{2}(\kappa - 1) \nabla_\mu \nabla_\nu - R_{\mu\nu} \right\} \delta(x, y) . \qquad (4.57)$$

It is obvious that for the operator Δ, (4.15), to be non-degenerate in the flat space-time it is necessary to choose the operator H as a second order differential operator

$$H^{\mu\nu} = \frac{1}{4\alpha^2} g^{-1/2} \left\{ -g^{\mu\nu} \Box + \beta \nabla^\mu \nabla^\nu + R^{\mu\nu} + P^{\mu\nu} \right\} \delta(x, y) , \qquad (4.58)$$

where α and β are numerical constants, $P^{\mu\nu}$ is an arbitrary symmetric tensor, e.g.

$$P^{\mu\nu} = p_1 R^{\mu\nu} + g^{\mu\nu} \left(p_2 R + p_3 \frac{1}{k^2} \right) , \qquad (4.59)$$

p_1, p_2 and p_3 being some arbitrary numerical constants. Such form of the operator H does not increase the order of the operator Δ, (4.15), and preserves its locality. Thus we obtain a very wide class of gauges. It involves six arbitrary parameters (κ, α, β, p_1, p_2, p_3). In particular, the harmonic De Donder–Fock–Landau gauge,

$$\nabla_\nu \left(h^\nu_\mu - \frac{1}{2} \delta^\nu_\mu h \right) = 0 , \qquad (4.60)$$

corresponds to $\kappa = 1$ and $\alpha = 0$. For $\alpha = 0$ the dependence on other parameters disappear.

It is most convenient to choose the "minimal" gauge

$$\alpha^2 = \alpha_0^2 \equiv f^2 , \qquad \beta = \beta_0 \equiv \frac{1}{3}(1 - 2\omega) , \qquad \kappa = \kappa_0 \equiv \frac{3\omega}{1+\omega} , \qquad (4.61)$$

which makes the operator Δ, (4.15), diagonal in leading derivatives, i.e., it takes the form

$$\Delta_{ik} = \frac{1}{2f^2} \left\{ \hat{E}_{(0)} \Box^2 + \nabla^\rho \hat{D}_{\rho\sigma} \nabla^\sigma + \nabla^\rho \hat{V}_\rho + \hat{V}_\rho \nabla^\rho + \hat{P} \right\} \delta(x, y) , \qquad (4.62)$$

where

$$\hat{E}_{(0)} = E^{\mu\nu,\alpha\beta}(\kappa = \kappa_0)\,, \qquad \hat{D}_{\rho\sigma} = D_{\rho\sigma}^{\ \mu\nu,\alpha\beta}\,,$$
$$\hat{V}_\rho = V_\rho^{\ \mu\nu,\alpha\beta}\,, \qquad \hat{P} = P^{\mu\nu,\alpha\beta}\,,$$

are some tensor matrices that satisfy the same symmetry conditions, (4.54). They have the following form

$$D_{\rho\sigma}^{\mu\nu,\alpha\beta} = \frac{1}{2f^2}\Bigg\{ 2R_{\rho\sigma}\left[g^{\alpha(\mu}g^{\nu)\beta} - \frac{1}{2}\left(1 + \frac{p_1}{16}\left(\frac{1+4\omega}{1+\omega}\right)^2\right) g^{\mu\nu}g^{\alpha\beta}\right]$$
$$+2\left(1 + \frac{p_1}{4}\left(\frac{1+4\omega}{1+\omega}\right)\right)\left(g^{\alpha\beta}R^{(\mu}_{(\rho}\delta^{\nu)}_{\sigma)} + g^{\mu\nu}R^{(\alpha}_{(\rho}\delta^{\beta)}_{\sigma)}\right)$$
$$+4g_{\rho\sigma}R^{(\mu|\alpha|\nu)\beta} - 2p_1\delta^{(\mu}_{(\rho}R^{\nu)(\alpha}\delta^{\beta)}_{\sigma)}$$
$$-4\left(\delta^{(\alpha}_{(\rho}g^{\beta)(\nu}R^{\mu)}_{\sigma)} + \delta^{(\mu}_{(\sigma}g^{\nu)(\beta}R^{\alpha)}_{\rho)}\right)$$
$$+\frac{4}{3}(1+\omega)\left[R^{\mu\nu}\left(\delta^{\alpha\beta}_{\rho\sigma} - g^{\alpha\beta}g_{\rho\sigma}\right) + R^{\alpha\beta}\left(\delta^{\mu\nu}_{\rho\sigma} - g^{\mu\nu}g_{\rho\sigma}\right)\right]$$
$$+\left(m_2^2 + \frac{2}{3}(1+\omega)R\right)\Big(-g_{\rho\sigma}g^{\alpha(\mu}g^{\nu)\beta} + g_{\rho\sigma}g^{\mu\nu}g^{\alpha\beta} - g^{\mu\nu}\delta^{\alpha\beta}_{\rho\sigma}$$
$$-\delta^{\mu\nu}_{\rho\sigma}g^{\alpha\beta} + 2\delta^{(\beta}_{(\sigma}g^{\alpha)(\mu}\delta^{\nu)}_{\rho)}\Big) + \left(p_2 R + p_3\frac{1}{k^2}\right)\Bigg[-2\delta^{(\beta}_{(\sigma}g^{\alpha)(\mu}\delta^{\nu)}_{\rho)}$$
$$+\frac{1}{2}\cdot\frac{1+4\omega}{1+\omega}\left(g^{\mu\nu}\delta^{\alpha\beta}_{\rho\sigma} + g^{\alpha\beta}\delta^{\mu\nu}_{\rho\sigma}\right)$$
$$-\frac{1}{8}\left(\frac{1+4\omega}{1+\omega}\right)^2 g_{\rho\sigma}g^{\mu\nu}g^{\alpha\beta}\Bigg]\Bigg\}\,,$$

$$V_\rho^{\mu\nu,\alpha\beta} = \frac{1}{2f^2}\Bigg\{\left(\frac{1}{2} + \frac{p_1}{8}\cdot\frac{1+4\omega}{1+\omega}\right)\left(g^{\alpha\beta}\nabla^{(\mu}R^{\nu)}_\rho - g^{\mu\nu}\nabla^{(\alpha}R^{\beta)}_\rho\right)$$
$$+\frac{1}{3}(1-2\omega)\left(g^{\mu\nu}\nabla_\rho R^{\alpha\beta} - g^{\alpha\beta}\nabla_\rho R^{\mu\nu}\right)$$
$$+g^{(\beta(\mu}\nabla^{\nu)}R^{\alpha)}_\rho - g^{(\mu(\beta}\nabla^{\alpha)}R^{\nu)}_\rho$$
$$+\frac{1}{12}(1+4\omega)\left(1 - \frac{3}{4}\cdot\frac{p_1}{1+\omega}\right)\left(g^{\alpha\beta}\delta^{(\mu}_\rho\nabla^{\nu)}R - g^{\mu\nu}\delta^{(\alpha}_\rho\nabla^{\beta)}R\right)$$

$$+\left(\frac{1-2\omega}{6}-\frac{p_2}{2}\right)\left(\delta_\rho^{(\alpha}g^{\beta)(\mu}\nabla^{\nu)}R-\delta_\rho^{(\mu}g^{\nu)(\alpha}\nabla^{\beta)}R\right)$$

$$+\frac{2}{3}(2-\omega)\left(\delta_\rho^{(\alpha}\nabla^{\beta)}R^{\mu\nu}-\delta_\rho^{(\mu}\nabla^{\nu)}R^{\alpha\beta}\right)$$

$$+\left(2+\frac{p_1}{2}\right)\left(\nabla^{(\alpha}R^{\beta)(\mu}\delta_\rho^{\nu)}-\nabla^{(\mu}R^{\nu)(\alpha}\delta_\rho^{\beta)}\right)\Bigg\}\,,$$

$$P^{\mu\nu,\alpha\beta}=\frac{1}{2f^2}\Bigg\{-\frac{1}{2}\left(1+\frac{p_1}{4}\cdot\frac{1+4\omega}{1+\omega}\right)\left(g^{\alpha\beta}R^\mu_\rho R^{\nu\rho}+g^{\mu\nu}R^\alpha_\rho R^{\beta\rho}\right)$$

$$-\left(g^{\mu(\alpha}g^{\beta)\nu}-\frac{1}{2}g^{\mu\nu}g^{\alpha\beta}\right)\Bigg[R_{\rho\sigma}R^{\rho\sigma}-\frac{1}{3}(1+\omega)R^2$$

$$+\frac{1}{3}(1+4\omega)\,\Box\,R-m_2^2(R-2\Lambda)\Bigg]$$

$$-\frac{3}{2}\left(1-\frac{p_1}{12}\cdot\frac{1+4\omega}{1+\omega}\right)R^{\rho\sigma}\left(g^{\alpha\beta}R^{\mu\ \nu}_{\ \rho\ \sigma}+g^{\mu\nu}R^{\alpha\ \beta}_{\ \rho\ \sigma}\right)$$

$$+(2+p_1)R^{\alpha(\mu}R^{\nu)\beta}-\frac{1}{2}p_1\left(R^{\rho(\mu}R^{\nu)(\alpha\ \beta)}_{\quad\ \rho}+R^{\rho(\alpha}R^{\beta)(\mu\ \nu)}_{\quad\ \rho}\right)$$

$$+6g^{(\beta(\nu}R^{\mu)\ \alpha)}_{\ \rho\ \sigma}R^{\rho\sigma}-\frac{4}{3}(1+\omega)R^{\mu\nu}R^{\alpha\beta}+4R^{(\mu\ \nu)\sigma}_{\ \rho}R^{\rho(\alpha\ \beta)}_{\quad\ \sigma}$$

$$+\left(m_2^2+\frac{2}{3}(1+\omega)R\right)\Bigg(R^{\mu\nu}g^{\alpha\beta}+R^{\alpha\beta}g^{\mu\nu}$$

$$-3g^{(\alpha(\mu}R^{\nu)\beta)}-R^{(\mu|\alpha|\nu)\beta}\Bigg)$$

$$+\left(p_2R+p_3\frac{1}{k^2}\right)\left(g^{(\alpha(\mu}R^{\nu)\beta)}-R^{(\mu|\alpha|\nu)\beta}\right)+2\,\Box\,R^{(\mu|\alpha|\nu)\beta}$$

$$-\omega\left(g^{\alpha\beta}\nabla^\mu\nabla^\nu R+g^{\mu\nu}\nabla^\alpha\nabla^\beta R\right)+2\,\Box\,R^{(\mu(\alpha}g^{\beta)\nu)}$$

$$+\frac{2}{3}(1+\omega)\Big(-g^{\alpha\beta}\,\Box\,R^{\mu\nu}-g^{\mu\nu}\,\Box\,R^{\alpha\beta}+\nabla^{(\alpha}\nabla^{\beta)}R^{\mu\nu}$$

$$+\nabla^{(\mu}\nabla^{\nu)}R^{\alpha\beta}\Big) - \frac{4}{3}(1-2\omega)g^{(\beta(\nu}\nabla^{\mu)}\nabla^{\alpha)}R\Big\} . \tag{4.63}$$

The divergences of the determinants of the operators H, (4.58), F, (4.56), and Δ, (4.62), can be calculated by means of the algorithms for the non-minimal vector operator of second order and the minimal tensor operator of forth order. These algorithms were obtained first in [107, 108, 109] and confirmed in [34] by using the generalized Schwinger–De Witt technique. In the dimensional regularization up to the terms $\sim \Box R$ they have the form

$$\log\det\{-\Box\delta^\mu_\nu + \beta\nabla^\mu\nabla_\nu + R^\mu_\nu + P^\mu_\nu\}\Big|^{\text{div}}$$

$$= \mathrm{i}\frac{2}{(n-4)(4\pi)^2}\int \mathrm{d}^4x\, g^{1/2}\Big\{-\frac{8}{45}R^*R^* + \frac{7}{60}C^2 + \frac{1}{36}R^2$$

$$+\frac{1}{6}(\xi+6)R_{\mu\nu}P^{\mu\nu} - \frac{1}{12}(\xi+2)RP + \frac{1}{48}\xi^2P^2$$

$$+\frac{1}{24}(\xi^2+6\xi+12)P_{\mu\nu}P^{\mu\nu}\Big\} , \tag{4.64}$$

where

$$\xi \equiv \frac{\beta}{1-\beta} , \qquad P \equiv P^\mu_{\ \mu} ,$$

and

$$\log\det\Big\{\hat{E}\Box^2 + \nabla^\mu\hat{D}_{\mu\nu}\nabla^\nu + \nabla^\mu\hat{V}_\mu + \hat{V}_\mu\nabla^\mu + \hat{P}\Big\}\Big|^{\text{div}}$$

$$= \mathrm{i}\frac{2}{(n-4)(4\pi)^2}\int \mathrm{d}^4x\, g^{1/2}\mathrm{tr}\Big\{\hat{1}\,\Big(-\frac{1}{180}R^*R^* + \frac{1}{60}C^2 + \frac{1}{36}R^2\Big)$$

$$+\frac{1}{6}\hat{\mathcal{R}}_{\mu\nu}\hat{\mathcal{R}}^{\mu\nu} - \hat{E}^{-1}\hat{P} + \frac{1}{12}R\hat{E}^{-1}\hat{D} - \frac{1}{6}R_{\mu\nu}\hat{E}^{-1}\hat{D}^{\mu\nu}$$

$$+\frac{1}{48}\hat{E}^{-1}\hat{D}\hat{E}^{-1}\hat{D} + \frac{1}{24}\hat{E}^{-1}\hat{D}_{\mu\nu}\hat{E}^{-1}\hat{D}^{\mu\nu}\Big\} , \tag{4.65}$$

where

$$\hat{D} \equiv \hat{D}^\mu_{\ \mu} ,$$

$\hat{E}$, $\hat{D}^{\mu\nu}$, $\hat{V}^\mu$ and $\hat{P}$ are the tensor matrices ($\hat{E} = E^{AB}$, $\hat{E}^{-1} = E^{-1}_{AB}$, $\hat{D}^{\mu\nu} = D^{AB\mu\nu}$ etc.), $\hat{1} = \delta^A_B$, $\hat{\mathcal{R}}_{\mu\nu} = \mathcal{R}^A_{\ B\mu\nu}$ is the commutator of covariant derivatives of the tensor field

$$[\nabla_\mu, \nabla_\nu] h^A = \mathcal{R}^A{}_{B\mu\nu} h^B \,, \tag{4.66}$$

'tr' means the matrix trace and n is the dimension of the space-time.

In our case, $h^A = h_{\mu\nu}$,

$$\mathcal{R}^A{}_{B\mu\nu} = \mathcal{R}_{\gamma\delta}{}^{\alpha\beta}{}_{,\mu\nu} = -2\delta^{(\alpha}_{(\gamma} R^{\beta)}{}_{\delta)\mu\nu} \,. \tag{4.67}$$

Using (4.64) and (4.65) we obtain in the minimal gauge (4.61)

$$\begin{aligned}\log\det H\Big|^{\mathrm{div}} &= \mathrm{i}\frac{2}{(n-4)(4\pi)^2}\int \mathrm{d}^4x\, g^{1/2}\Big\{-\frac{8}{45}R^*R^* + \frac{7}{60}C^2 \\ &+\frac{1}{36}R^2 + \frac{13+10\omega}{12(1+\omega)}R_{\mu\nu}P^{\mu\nu} \\ &-\frac{5+2\omega}{24(1+\omega)}RP + \frac{1}{192}\left(\frac{1-2\omega}{1+\omega}\right)^2 P^2 \\ &+\frac{28\omega^2+80\omega+61}{96(\omega+1)^2}P_{\mu\nu}P^{\mu\nu}\Big\}\,,\end{aligned} \tag{4.68}$$

$$\begin{aligned}\log\det F\Big|^{\mathrm{div}} &= \mathrm{i}\frac{2}{(n-4)(4\pi)^2}\int \mathrm{d}^4x\, g^{1/2} \\ &\times\Big\{-\frac{1}{540}(20\omega^2+100\omega+41)R^*R^* \\ &+\frac{1}{135}(5\omega^2+25\omega+2)C^2 \\ &+\frac{1}{81}(5\omega^2+16\omega+20)R^2\Big\}\,,\end{aligned} \tag{4.69}$$

$$\begin{aligned}\log\det \Delta\Big|^{\mathrm{div}} &= \mathrm{i}\frac{2}{(n-4)(4\pi)^2}\int \mathrm{d}^4x\, g^{1/2}\Big\{-\frac{1}{54}(4\omega^2+20\omega+253)R^*R^* \\ &+\frac{1}{54}(4\omega^2+20\omega+367)C^2 \\ &+\frac{1}{162}(200\omega^2+334\omega+107)R^2\end{aligned}$$

$$+\frac{1}{6}\left(40\omega-26-\frac{3}{\omega}\right)\frac{f^2}{k^2}R$$

$$+\frac{1}{k^4}\left[\frac{4}{3}\lambda(14f^2+\nu^2)+\frac{1}{2}(5f^4+\nu^4)\right]$$

$$-\frac{13+10\omega}{12(1+\omega)}R_{\mu\nu}P^{\mu\nu}+\frac{5+2\omega}{24(1+\omega)}RP$$

$$-\frac{1}{192}\left(\frac{1-2\omega}{1+\omega}\right)^2P^2-\frac{28\omega^2+80\omega+61}{96(\omega+1)^2}P_{\mu\nu}P^{\mu\nu}\Bigg\}\,. \tag{4.70}$$

Substituting the obtained expressions (4.68)–(4.70) in (4.52) we get the divergences of the standard one-loop effective action off mass shell

$$\Gamma^{\rm div}_{(1)}=-\frac{1}{(n-4)(4\pi)^2}\int d^4x\, g^{1/2}\Bigg\{\beta_1R^*R^*+\beta_2C^2$$

$$+\beta_3R^2+\beta_4\frac{1}{k^4}+\gamma\frac{1}{k^2}(R-4\Lambda)\Bigg\}\,, \tag{4.71}$$

where

$$\beta_1=-\frac{196}{45}\,,$$

$$\beta_2=\frac{133}{20}\,, \tag{4.72}$$

$$\beta_3=\frac{5}{18}\frac{f^4}{\nu^4}+\frac{5}{6}\frac{f^2}{\nu^2}+\frac{5}{36}\,, \tag{4.73}$$

$$\beta_4=\frac{1}{2}(5f^4+\nu^4)+\frac{2}{3}\lambda\left(10\frac{f^4}{\nu^2}+15f^2-\nu^2\right)\,, \tag{4.74}$$

$$\gamma=\frac{5}{3}\frac{f^4}{\nu^2}-\frac{13}{6}f^2-\frac{1}{2}\nu^2\,. \tag{4.75}$$

Therefrom it is immediately seen that the gauge fixing tensor, $P^{\mu\nu}$, (4.59) does not enter the result. In the next section we will calculate the divergences of the effective action in arbitrary gauge and will show that the tensor $P^{\mu\nu}$ does not contribute in the divergences in general case too. If one puts $P_{\mu\nu}=0$ then the divergences of the operator H do not depend on the gauge fixing parameters at all.

Our result for divergences, (4.71)–(4.75), does not coincide with the results of the papers [107, 108, 109, 111] in the coefficient β_3, (4.73). Namely, the last term in (4.73) is equal to $5/36$ instead of the incorrect value $-1/36$

obtained in [107, 108, 109, 111]. We will check our result, (4.73), by means of completely independent computation on the De Sitter background in Sect. 4.5.

4.3 One-Loop Divergences in Arbitrary Gauge and Vilkovisky's Effective Action

Let us study now the dependence of the obtained result for the divergences of the standard effective action, (4.71)–(4.75), on the choice of the gauge. Let us consider the variation of the one-loop effective action (4.1) with respect to variation of the gauge condition (i.e., the functions χ_μ and $H^{\mu\nu}$)

$$\delta\Gamma_{(1)} = -\frac{1}{\mathrm{i}}\left\{\left(\Delta^{-1\,ik}\chi_{\mu k}H^{\mu\nu} - R^i{}_\alpha F^{-1\,\alpha\nu}\right)\delta\chi_{\nu i}\right.$$

$$\left. +\frac{1}{2}\left(\chi_{\mu i}\Delta^{-1\,ik}\chi_{\nu k} - H^{-1}_{\mu\nu}\right)\delta H^{\mu\nu}\right\}. \qquad (4.76)$$

Using the Ward identities

$$\Delta^{-1\,ik}\chi_{\mu k}H^{\mu\nu} - R^i{}_\alpha F^{-1\,\alpha\nu} = -\Delta^{-1\,ik}\varepsilon_j R^j{}_{\alpha,k}F^{-1\,\alpha\nu}\,, \qquad (4.77)$$

$$\chi_{\mu i}\Delta^{-1\,ik}\chi_{\nu k} - H^{-1}_{\mu\nu} = -H^{-1}_{\mu\alpha}F^{-1\,\beta\alpha}\varepsilon_j R^j{}_{\beta,k}$$

$$\times\left(R^k{}_\gamma - \Delta^{-1\,kn}\varepsilon_m R^m{}_{\gamma,n}\right)F^{-1\,\gamma\delta}H^{-1}_{\delta\nu}\,, \qquad (4.78)$$

that follow from the Nöther identity, (4.6), one can derive from (4.76)

$$\delta\Gamma_{(1)} = -\frac{1}{\mathrm{i}}\left\{-\Delta^{-1\,ik}\varepsilon_j R^j{}_{\alpha,k}F^{-1\,\alpha\nu}\delta\chi_{\nu i}\right.$$

$$\left. +\frac{1}{2}F^{-1\,\beta\alpha}\varepsilon_j R^j{}_{\beta,k}\left(R^k{}_\gamma - \Delta^{-1\,kn}\varepsilon_m R^m{}_{\gamma,n}\right)F^{-1\,\gamma\delta}\delta(H^{-1}_{\delta\alpha})\right\}. \qquad (4.79)$$

From here, it follows, in particular, that the one-loop effective action on mass shell, $\varepsilon = 0$, (4.5), does not depend on the gauge,

$$\delta\Gamma\Big|_{\text{on-shell}} = 0. \qquad (4.80)$$

Since the effective action on the mass shell is well defined, it is analytical in background fields in the neighborhood of the mass shell (4.5). Therefore, it can be expanded in powers of the extremal [223]. As the extremal has the background dimension (in our case, (4.50), equal to four in mass units), this expansion will be, in fact, an expansion in the background dimension. It is obvious, that to calculate the divergences of the effective action it is sufficient to limit oneself to the terms of background dimension not greater than four. Thus one can obtain the divergences by taking into account only *linear* terms in the extremal. Moreover, from the dimensional grounds it follows that only the trace of the extremal (4.50),

$$\varepsilon \equiv g_{\mu\nu}\varepsilon^{\mu\nu} = g^{1/2}\left\{\frac{1}{k^2}(R-4\Lambda) - \frac{1}{\nu^2}\Box R\right\}, \tag{4.81}$$

contributes to the divergences. Therefore, only γ-coefficient, (4.75), in the divergent part of the effective action depends on the gauge parameters. The β-functions, (4.72)–(4.74), do not depend on the gauge.

So, from (4.79) we obtain the variation of the one-loop effective action with respect to the variation of the gauge

$$\delta\Gamma^{\mathrm{div}}_{(1)} = \frac{1}{\mathrm{i}}\left\{\varepsilon_j R^j_{\ \alpha,k}\Delta^{-1\ ki}F^{-1\ \alpha\beta}\delta\chi_{\beta i} - \frac{1}{2}\varepsilon_j R^j_{\ \alpha,k}R^k_{\ \beta}F^{-1\ \beta\gamma}F^{-1\ \alpha\delta}\delta(H^{-1}_{\ \gamma\delta})\right\}\Bigg|^{\mathrm{div}}. \tag{4.82}$$

From here one can obtain the divergences of the effective action in any gauge. To do this one has, first, to fix some form of the gauge condition with arbitrary parameters, then, to calculate the divergences of the effective action for some convenient choice of the gauge parameters and, finally, to integrate the equation (4.82) over the gauge parameters.

Let us restrict ourselves to the covariant De Witt gauge, (4.56), with arbitrary gauge parameters α, β and $P^{\mu\nu}$. Since the coefficient at the variation δH^{-1} in (4.82) does not depend on the operator H (i.e., on the gauge parameters α, β and $P^{\mu\nu}$), one can integrate over the operator H immediately. Thus we obtain the divergences of the effective action in arbitrary gauge

$$\Gamma^{\mathrm{div}}_{(1)}(\kappa,\alpha,\beta,P) = \Gamma^{\mathrm{div}}_{(1)}(\kappa_0,\alpha_0,\beta_0) + \frac{1}{\mathrm{i}}\left\{\int\limits_{\kappa_0}^{\kappa}\mathrm{d}\kappa\; U^{\mathrm{div}}_1(\kappa) - \frac{1}{2}\left[U^{\mathrm{div}}_2(\kappa,\alpha,\beta,P) - U^{\mathrm{div}}_2(\kappa,\alpha_0,\beta_0,P)\right]\right\}, \tag{4.83}$$

where $\Gamma^{\mathrm{div}}_{(1)}(\kappa_0,\alpha_0,\beta_0)$ is the divergent part of the effective action in the minimal gauge, (4.71)–(4.75),

$$U_1(\kappa) = \varepsilon_j R^j{}_{\alpha,k} \Delta^{-1\,ki}(\kappa, \alpha_0, \beta_0, P) F^{-1\,\alpha\beta}(\kappa) R^n{}_\beta E'_{in} \,, \tag{4.84}$$

$$E'_{in} = \frac{d}{d\kappa} E_{in} = -\frac{1}{4} g^{1/2} g^{\mu\nu} g^{\alpha\beta} \delta(x,y) \,,$$

$$U_2(\kappa, \alpha, \beta, P) = \varepsilon_j R^j{}_{\alpha,k} R^k{}_\beta F^{-1\,\beta\gamma}(\kappa) F^{-1\,\alpha\delta}(\kappa) H^{-1}_{\gamma\delta}(\alpha, \beta, P) \Big|^{\mathrm{div}} \,. \tag{4.85}$$

To calculate the quantities U_1, (4.84), and U_2, (4.85), one has to find the ghost propagators $F^{-1}(\kappa)$ and $H^{-1}_{\gamma\delta}(\alpha, \beta, P)$ for arbitrary κ, α, β and P and the gravitational propagator $\Delta^{-1}(\kappa, \alpha_0, \beta_0)$ for arbitrary parameter κ and minimal values of other parameters, α_0 and β_0. The whole background dimension that causes the divergences is contained in the extremal ε_i. So, when calculating the divergences of the quantities U_1, (4.84), and U_2, (4.85), one can take all propagators to be free, i.e., one can neglect the background quantities, like the space-time curvature, the commutator of covariant derivatives etc., and the mass terms. This is so because together with any dimensional terms there appear automatically a Green function $\Box^{-1}$, which makes the whole term finite. Therefore, in particular, the gauge fixing tensor $P^{\mu\nu}$ does not contribute to the divergences of the effective action at all. For the minimal gauge, (4.61), we have shown this in previous section by an explicit calculation.

Using the explicit forms of the operators $F(\kappa)$, (4.56), $H(\alpha, \beta)$, (4.58), and $\Delta(\kappa, \alpha_0, \beta_0)$, (4.15), (4.53), we find the free Green functions of these operators

$$F^{-1\,\mu\nu}(\kappa) = \frac{1}{2}\left(-g^{\mu\nu}\Box + \frac{\kappa-1}{\kappa-3}\nabla^\mu\nabla^\nu\right)\Box^{-2} g^{-1/2}\delta(x,y) \,, \tag{4.86}$$

$$H^{-1}_{\mu\nu}(\alpha, \beta) = 4\alpha^2\left(-g_{\mu\nu}\Box - \frac{\beta}{1-\beta}\nabla_\mu\nabla_\nu\right)\Box^{-2} g^{1/2}\delta(x,y) \,, \tag{4.87}$$

$$\Delta^{-1\,ik}(\kappa, \alpha_0, \beta_0) = 2f^2\Bigg\{E^{-1}_{(0)\mu\nu,\alpha\beta}\Box^2 + \rho_2\nabla_\mu\nabla_\nu\nabla_\alpha\nabla_\beta$$
$$+\rho_1\left(g_{\mu\nu}\nabla_\alpha\nabla_\beta + g_{\alpha\beta}\nabla_\mu\nabla_\nu\right)\Box\Bigg\}\Box^{-4} g^{-1/2}\delta(x,y) \,, \tag{4.88}$$

where

$$E^{-1}_{(0)\mu\nu,\alpha\beta} \equiv E^{-1}_{\mu\nu,\alpha\beta}\Big|_{\kappa=\kappa_0} = g_{\mu(\alpha}g_{\beta)\nu} - \frac{1+4\omega}{12\omega}g_{\mu\nu}g_{\alpha\beta} \,, \tag{4.89}$$

$$\rho_1 = \frac{1}{3\omega}\left(\omega + 1 + \frac{3}{\kappa-3}\right) \,, \tag{4.90}$$

$$\rho_2 = -\frac{4}{3\omega(\omega+1)}\left(\omega+1+\frac{3}{\kappa-3}\right)^2 . \tag{4.91}$$

Let us note that in the minimal gauge, (4.61), $\rho_1 = \rho_2 = 0$ and, therefore, $\Delta^{-1\;ik}_{(0)} = 2f^2 E^{-1}_{(0)\mu\nu,\alpha\beta}\,\Box^{-2} g^{-1/2}\delta(x,y)$.

Substituting the free propagators, (4.86)–(4.88), in (4.84) and (4.85) we obtain the divergences of the quantities U_1 and U_2

$$U_1^{\mathrm{div}} = \frac{f^2}{\kappa-3}\int \mathrm{d}^4x\, \varepsilon^{\mu\nu}\left\{(\rho_2+6\rho_1-2\kappa_0^{-1})\nabla_\mu\nabla_\nu\Box^{-3}g^{-1/2}\delta(x,y)\Big|_{y=x}^{\mathrm{div}}\right.$$

$$\left. -\left(\rho_1-\frac{1+\omega}{3\omega}\right) g_{\mu\nu}\Box^{-2}g^{-1/2}\delta(x,y)\Big|_{y=x}^{\mathrm{div}}\right\} , \tag{4.92}$$

$$U_2^{\mathrm{div}} = 2\alpha^2\left(3+\frac{4}{(\kappa-3)^2(1-\beta)}\right)\int \mathrm{d}^4x\, \varepsilon^{\mu\nu}\nabla_\mu\nabla_\nu\Box^{-3}g^{-1/2}\delta(x,y)\Big|_{y=x}^{\mathrm{div}} . \tag{4.93}$$

Using the divergences of the coincidence limits of the Green functions and their derivatives in the dimensional regularization

$$\Box^{-2}g^{-1/2}\delta(x,y)\Big|_{y=x}^{\mathrm{div}} = -\mathrm{i}\frac{2}{(n-4)(4\pi)^2} ,$$

$$\nabla_\mu\nabla_\nu\Box^{-3}g^{-1/2}\delta(x,y)\Big|_{y=x}^{\mathrm{div}} = -\mathrm{i}\frac{2}{(n-4)(4\pi)^2}\frac{1}{4}g_{\mu\nu} , \tag{4.94}$$

and the explicit form of the extremal $\varepsilon^{\mu\nu}$, (4.50), we obtain

$$U_1^{\mathrm{div}} = \mathrm{i}\frac{1}{(n-4)(4\pi)^2}\frac{6\nu^2}{(\kappa-3)^2}\left(1+\frac{2}{(\omega+1)(\kappa-3)}\right)\int \mathrm{d}^4x\, g^{1/2}\frac{1}{k^2}(R-4\Lambda) ,$$

$$U_2^{\mathrm{div}} = -\mathrm{i}\frac{1}{(n-4)(4\pi)^2}\alpha^2\left(3+\frac{4}{(\kappa-3)^2(1-\beta)}\right)\int \mathrm{d}^4x\, g^{1/2}\frac{1}{k^2}(R-4\Lambda) . \tag{4.95}$$

Substituting these expressions in (4.83) and integrating over κ we obtain finally

$$\Gamma^{\mathrm{div}}_{(1)}(\kappa,\alpha,\beta,P) = \Gamma^{\mathrm{div}}_{(1)}(\kappa_0,\alpha_0,\beta_0)$$

$$-\frac{1}{(n-4)(4\pi)^2}\int \mathrm{d}^4x\, g^{1/2}\Delta\gamma(\kappa,\alpha,\beta)\frac{1}{k^2}(R-4\Lambda) ,$$

where

$$\Delta\gamma(\kappa,\alpha,\beta) = \frac{13}{6}f^2 + \frac{4}{3}\nu^2 - \frac{3}{2}\alpha^2 - \frac{2\alpha^2}{(\kappa-3)^2(1-\beta)} + \frac{6\nu^2(\kappa-2)}{(\kappa-3)^2} \,. \quad (4.96)$$

Thus the off-shell divergences of the effective action in arbitrary gauge have the same form, (4.71), where the coefficients β_1, β_2, β_3 and β_4 do not depend on the gauge and are given by the expressions (4.72)–(4.74), and the γ-coefficient reads

$$\gamma(\kappa,\alpha,\beta) = \frac{5}{3}\frac{f^4}{\nu^2} + \frac{5}{6}\nu^2 - \frac{3}{2}\alpha^2 - \frac{2\alpha^2}{(\kappa-3)^2(1-\beta)} + \frac{6\nu^2(\kappa-2)}{(\kappa-3)^2} \,. \quad (4.97)$$

In particular, in the harmonic gauge, (4.60), ($\kappa = 1$ and $\alpha = 0$), we have

$$\gamma(1,0,\beta) = \frac{5}{3}\frac{f^4}{\nu^2} - \frac{2}{3}\nu^2 \,. \quad (4.98)$$

The dependence of the divergences on the parametrization of the quantum field also exhibits only in the γ-coefficient. Rather than to study this dependence, let us calculate the divergences of the Vilkovisky's effective action $\tilde{\Gamma}$, (4.24), that does not depend neither on the gauge nor on the parametrization of the quantum field. From (4.24) we have

$$\tilde{\Gamma}^{\text{div}}_{(1)} = -\frac{1}{2i}\left(\log\det\tilde{\Delta}\Big|^{\text{div}} - \log\det H\Big|^{\text{div}} - 2\log\det F\Big|^{\text{div}}\right) \,. \quad (4.99)$$

The Vilkovisky's effective action $\tilde{\Gamma}$, (4.24), (4.99), differs from the usual one, (4.14), (4.52), only by the operator $\tilde{\Delta}$, (4.25). It is obtained from the operator Δ, (4.15), by substituting the covariant functional derivatives instead of the usual ones:

$$S_{,ik} \to \nabla_i\nabla_k S = \mathcal{D}_i\mathcal{D}_k S - T^j{}_{ik}\varepsilon_j = S_{,ik} - \Gamma^j{}_{ik}\varepsilon_j \,,$$

$$\tilde{\Delta}_{ik} = \tilde{\Delta}^{\text{loc}}_{ik} + T^j{}_{ik}\varepsilon_j = \Delta_{ik} + \Gamma^j{}_{ik}\varepsilon_j \,, \quad (4.100)$$

where

$$\tilde{\Delta}^{\text{loc}}_{ik} = -\mathcal{D}_i\mathcal{D}_k S + \chi_{\mu i}H^{\mu\nu}\chi_{\nu k} \,. \quad (4.101)$$

Since the non-metric part of the connection $T^j{}_{ik}$, (4.34), is non-local, the operator $\tilde{\Delta}$, (4.100), is an integro-differential one. The calculation of the determinants of such operators offers a serious problem. However, as the non-local part of the operator $\tilde{\Delta}$, (4.100), is proportional to the extremal ε_i, it exhibits only off mass shell. Therefore, the calculation of the determinant of the operator $\tilde{\Delta}$, (4.100), can be based on the expansion in the non-local part, $T^j{}_{ik}\varepsilon_j$. To calculate the divergences it is again sufficient to limit oneself only to linear terms

$$\log\det\tilde{\Delta}\Big|^{\text{div}} = \log\det\tilde{\Delta}_{\text{loc}}\Big|^{\text{div}} + \tilde{\Delta}^{-1}_{\text{loc}}{}^{mn}T^i{}_{mn}\varepsilon_i\Big|^{\text{div}} \,. \quad (4.102)$$

To calculate this expression one can choose any gauge, because the answer for the Vilkovisky's effective action does not depend on the gauge. Let us choose the De Witt gauge, (4.56),

$$\chi_{\mu i} = R^k{}_\mu E_{ki} \,, \qquad F_{\mu\nu} = N_{\mu\nu} \,. \tag{4.103}$$

Using the Ward identities for the Green function of the operator $\tilde{\Delta}_{\text{loc}}$, (4.101), in De Witt gauge we get (up to terms proportional to the extremal)

$$\begin{aligned} B^\alpha_{;i} \tilde{\Delta}^{-1\,ik}_{\text{loc}} &= N^{-1\,\alpha\mu} H^{-1}_{\mu\nu} N^{-1\,\nu\beta} R^k{}_\beta + O(\varepsilon) \,, \\ B^\alpha_{;i} \tilde{\Delta}^{-1\,ik}_{\text{loc}} B^\beta{}_k &= N^{-1\,\alpha\mu} H^{-1}_{\mu\nu} N^{-1\,\nu\beta} + O(\varepsilon) \,. \end{aligned} \tag{4.104}$$

Using the explicit form of $T^i{}_{mn}$, (4.34), in (4.102) and (4.104) we obtain

$$\log\det \tilde{\Delta}\Big|^{\text{div}} = \log\det \tilde{\Delta}_{\text{loc}}\Big|^{\text{div}} - U_3^{\text{div}} \,,$$

where

$$U_3 = \varepsilon_j \mathcal{D}_k R^j{}_\alpha R^k{}_\beta N^{-1\,\alpha\mu} H^{-1}_{\mu\nu} N^{-1\,\nu\beta} \,. \tag{4.105}$$

Finally, one has to fix the operator H (i.e., the parameters α and β) and to determine the parameter of the metric of the configuration space, κ, (4.45), (4.46).

In the paper [223] some conditions on the metric E_{ik} were formulated, that make it possible to fix the parameter κ. First, the metric E_{ik} must be contained in the term with highest derivatives in the action $S(\varphi)$. Second, the operator $N_{\mu\nu}$, (4.28), (4.48), must be non-degenerate within the perturbation theory. To find the metric E_{ik}, (4.45), i.e., the parameter κ, (4.46), one should consider the second variation of the action on the physical quantum fields, $h_\perp = \Pi_\perp h$, that satisfy the De Witt gauge conditions, $R_{i\mu} h^i_\perp = 0$, and identify the metric with the matrix E in the highest derivatives

$$\begin{aligned} h^i_\perp(-S_{,ik}) h^k_\perp &= \frac{1}{2f^2} \int \mathrm{d}^4x\, h^\perp_{\mu\nu} g^{1/2} E^{\mu\nu,\alpha\beta}(\kappa) \,\Box^2 h^\perp_{\alpha\beta} \\ &\quad + \text{terms with the curvature} \,. \end{aligned} \tag{4.106}$$

This condition leads to a quadratic equation for κ that has two solutions

$$\kappa_1 = 3\frac{\omega}{\omega+1} \,, \qquad \kappa_2 = 3 \,. \tag{4.107}$$

As we already noted above, the value $\kappa = \kappa_2 = 3$ is unacceptable, since the operator N, (4.48), in this case is degenerate on the flat background. Therefore, we find finally

$$\kappa = \tilde{\kappa} \equiv 3\frac{\omega}{\omega+1} \,. \tag{4.108}$$

Let us note that this value of κ coincides with the minimal one, $\tilde{\kappa} = \kappa_0$, (4.61). Thus, if we choose the operator H in the same form, (4.58), with the minimal parameters, $\alpha = \alpha_0$ and $\beta = \beta_0$, (4.61), then the operator $\tilde{\Delta}_{\text{loc}}$ becomes a minimal operator of the form (4.62)

$$\tilde{\Delta}^{\text{loc}}_{ik} = \Delta_{ik} + \left\{ {}^{j}{}_{ik} \right\} \varepsilon_j \, , \tag{4.109}$$

where Δ_{ik} is given by the expressions (4.62) and (4.63).

The divergences of the determinant of the operator $\tilde{\Delta}_{\text{loc}}$, (4.109), can be calculated either by direct application of the algorithm (4.65) or by means of the expansion in the extremal

$$\log\det \tilde{\Delta}_{\text{loc}}\Big|^{\text{div}} = \log\det \Delta_{\text{loc}}\Big|^{\text{div}} + U_4^{\text{div}} \, , \tag{4.110}$$

where

$$U_4 = \Delta^{-1\, mn} \left\{ {}^{i}{}_{mn} \right\} \varepsilon_i .$$

Using the formulas (4.99), (4.105), (4.110) and (4.52) we obtain the divergences of the Vilkovisky's effective action

$$\tilde{\Gamma}^{\text{div}}_{(1)} = \Gamma^{\text{div}}_{(1)}(\kappa_0, \alpha_0, \beta_0) + \frac{1}{2\mathrm{i}}(U_3^{\text{div}} - U_4^{\text{div}}) \, , \tag{4.111}$$

where $\Gamma^{\text{div}}_{(1)}(\kappa_0, \alpha_0, \beta_0)$ are the divergences of the effective action in the minimal gauge, (4.71)–(4.75). The quantities U_3^{div} and U_4^{div} are calculated by using the free propagators, (4.86)-(4.88), in the minimal gauge, (4.61),

$$U_3^{\text{div}} = U_2^{\text{div}}(\kappa_0, \alpha_0, \beta_0) - 4f^2 \int \mathrm{d}^4x \, \varepsilon^{\mu\nu} \left\{ {}^{\alpha\beta,\rho\sigma}_{\mu\nu} \right\}$$

$$\times \nabla_\alpha \nabla_\rho \left(-g_{\beta\sigma} \Box + \frac{1}{3}(1 - 2\omega) \nabla_\beta \nabla_\sigma \right) \Box^{-4} g^{-1/2} \delta(x, y) \Bigg|^{\text{div}}_{y=x} \, ,$$

$$U_4^{\text{div}} = \int \mathrm{d}^4x \, \varepsilon^{\mu\nu} \left\{ {}^{\alpha\beta,\gamma\delta}_{\mu\nu} \right\} 2f^2 E^{-1}_{\alpha\beta,\gamma\delta} \Box^{-2} g^{-1/2} \delta(x, y) \Bigg|^{\text{div}}_{y=x} \, , \tag{4.112}$$

where $U_2^{\text{div}}(\kappa_0, \alpha_0, \beta_0)$ is given by the formula (4.95) in the minimal gauge (4.61). Using the divergences of the coincidence limits of the derivatives of the Green functions, (4.94), and the Christoffel connection, (4.47), and substituting the minimal values of the parameters κ, α and β, (4.61), we obtain

$$U_3^{\text{div}} = -\mathrm{i} \frac{1}{(n-4)(4\pi)^2} \frac{4}{3} (2\nu^2 + f^2) \int \mathrm{d}^4x \, g^{1/2} \frac{1}{k^2}(R - 4\Lambda) \, ,$$

$$U_4^{\text{div}} = -\mathrm{i} \frac{1}{(n-4)(4\pi)^2} 6(\nu^2 - f^2) \int \mathrm{d}^4x \, g^{1/2} \frac{1}{k^2}(R - 4\Lambda) \, . \tag{4.113}$$

Thus, the off-shell divergences of the one-loop effective action, $\tilde{\Gamma}_{(1)}$, have the standard form (4.71), where the β-functions are determined by the same expressions (4.72)–(4.74) and the γ-coefficient, (4.75), has an extra contribution due to the quantities $U_3^{\rm div}$ and $U_4^{\rm div}$, (4.113), in (4.111). It has the form

$$\tilde{\gamma} = \gamma(\kappa_0, \alpha_0, \beta_0) + \frac{1}{3}(11f^2 - 5\nu^2) = \frac{5}{3}\frac{f^4}{\nu^2} + \frac{3}{2}f^2 - \frac{13}{6}\nu^2 , \qquad (4.114)$$

where $\gamma(\kappa_0, \alpha_0, \beta_0)$ is given by (4.75).

4.4 Renormalization Group and Ultraviolet Asymptotics

The structure of the divergences of the effective action, (4.71), indicates that the higher-derivative quantum gravity is renormalizable off mass shell. Thus one can apply the renormalization group methods to study the high-energy behavior of the effective (running) coupling constants [50, 155, 226, 229]. The dimensionless constants ϵ, f^2, ν^2 and λ are the "essential" coupling constants [229] but the Einstein dimensional constant k^2 is "non-essential" because its variation can be compensated by a reparametrization of the quantum field, i.e., up to total derivatives

$$\left.\frac{\partial S}{\partial k^2}\right|_{\rm on-shell} = 0 . \qquad (4.115)$$

Using the one-loop divergences of the effective action, (4.71), we obtain in the standard way the renormalization group equations for the coupling constants of the renormalized effective action [50, 229]

$$\frac{d}{dt}\epsilon = \beta_1 , \qquad \frac{d}{dt}f^2 = -2\beta_2 f^4 , \qquad (4.116)$$

$$\frac{d}{dt}\nu^2 = 6\beta_3\nu^4 = \frac{5}{3}f^4 + 5f^2\nu^2 + \frac{5}{6}\nu^4 , \qquad (4.117)$$

$$\frac{d}{dt}\lambda = \frac{1}{2}\beta_4 = \frac{1}{4}(\nu^4 + 5f^4) + \frac{1}{3}\lambda\left(10\frac{f^4}{\nu^2} + 15f^2 - \nu^2\right) , \qquad (4.118)$$

$$\frac{d}{dt}k^2 = \gamma k^2 , \qquad (4.119)$$

where $t = (4\pi)^{-2}\log(\mu/\mu_0)$, μ is the renormalization parameter and μ_0 is a fixed energy scale.

The ultraviolet behavior of the essential coupling constants $\epsilon(t)$, $f^2(t)$, $\nu^2(t)$ and $\lambda(t)$ as $t \to \infty$ is determined by the coefficients (4.72)–(4.74). They play the role of generalized Gell-Mann–Low β-function, (1.47), and do not depend neither on the gauge condition nor on the parametrization of the quantum field. The non-essential coupling constant $k^2(t)$ is, in fact, simply a field renormalization constant. Thus the γ-coefficient, (4.75), (4.97) and (4.114), play in (4.119) the role of the anomalous dimension (1.48). Correspondingly, the ultraviolet behavior of the constant $k^2(t)$ depends essentially both on the gauge and the parametrization of the quantum field. It is obvious that one can choose the gauge condition in such a way that the coefficient γ, (4.97), is equal to zero, $\gamma = 0$. In this case the Einstein coupling constant is not renormalized at all, i.e., $k^2(t) = k^2(0) = \text{const}$.

The equations (4.116) can be easily solved

$$\epsilon(t) = \epsilon(0) + \beta_1 t,$$

$$f^2(t) = \frac{f^2(0)}{1 + 2\beta_2 f^2(0) t} \,. \tag{4.120}$$

Noting that $\beta_1 < 0$ and $\beta_2 > 0$, (4.72), we find the following. First, the topological coupling constant $\epsilon(t)$ becomes negative in the ultraviolet limit $(t \to \infty)$ and its absolute value grows logarithmically regardless of the initial value $\epsilon(0)$. Second, the Weyl coupling constant $f^2(t)$ is either asymptotically free (at $f^2 > 0$) or has a "zero-charge" singularity (at $f^2 < 0$). We limit ourselves to the first case, $f^2 > 0$, since, on the one hand, this condition ensures the stability of the flat background under the spin-2 tensor excitations, and, on the other hand, it leads to a positive contribution of the Weyl term to the Euclidean action (4.49).

The solution of the equation (4.117) can be written in the form

$$\nu^2(t) = \frac{c_1 f^{2p}(t) - c_2 f_*^{2p}}{f^{2p}(t) - f_*^{2p}} f^2(t) \,, \tag{4.121}$$

where

$$c_{1,2} = \frac{1}{50}(-549 \pm \sqrt{296401}) \approx \begin{cases} -0.091 \\ -21.87 \end{cases} , \tag{4.122}$$

$$p = \frac{\sqrt{296401}}{399} \approx 1.36 \,, \qquad f_*^{2p} \equiv \frac{\nu^2(0) - c_1 f^2(0)}{\nu^2(0) - c_2 f^2(0)} f^{2p}(0) \,. \tag{4.123}$$

There are also two special solutions

$$\nu_{1,2}^2(t) = c_{1,2} f^2(t) \,, \tag{4.124}$$

that correspond to the values $f_*^{2p} = 0, \infty$ in (4.121). These solutions are asymptotically free but only $\nu_2^2(t)$ is stable in the ultraviolet limit.

The behavior of the conformal coupling constant $\nu^2(t)$ depends essentially on its initial value $\nu^2(0)$. In the case $\nu^2(0) > c_1 f^2(0)$ we have $f^{2p}(0) > f_*^{2p} > 0$ and, therefore, the function $\nu^2(t)$, (4.121), has a typical "zero-charge" singularity at a finite scale $t = t_*$ determined from $f^{2p}(t_*) = f_*^{2p}$:

$$\nu^2(t)\Big|_{t\to t_*} = c_3 \frac{f_*^{2(p+1)}}{f^{2p}(t) - f_*^{2p}} + O(1) , \qquad (4.125)$$

where

$$c_3 = c_1 - c_2 = \frac{\sqrt{296401}}{25} \approx 21.78 . \qquad (4.126)$$

In the opposite case, $\nu^2(0) < c_1 f^2(0)$, the function $\nu^2(t)$, (4.121), does not have any singularities and is asymptotically free,

$$\nu^2(t)\Big|_{t\to\infty} = c_2 f^2(t) - c_3 f_*^{-2p} f^{2(p+1)}(t) + O\left(f^{2(1+2p)}\right) . \qquad (4.127)$$

Thus, contrary to the conclusions of the papers [107, 108, 109, 111], we find that in the region $\nu^2 > 0$ there are no asymptotically free solutions. The asymptotic freedom for the conformal coupling constant $\nu^2(t)$ can be achieved only in the negative region $\nu(0) < 0$, (4.127).

The exact solution of the equation for the dimensionless cosmological constant, (4.118), has the form

$$\lambda(t) = \Phi(t) \left\{ \Phi^{-1}(0)\lambda(0) + \int_0^t \mathrm{d}\tau A(\tau)\Phi^{-1}(\tau) \right\} , \qquad (4.128)$$

where

$$A(\tau) = \frac{1}{4}\left(5f^4(t) + \nu^4(t)\right) ,$$

$$\Phi(t) = \left|c_1 f^{2p}(t) - c_2 f_*^{2p}\right|^2 \left|f^{2p}(t) - f_*^{2p}\right|^{2/5} \left|f^2(t)\right|^{-q} , \qquad (4.129)$$

$$q = \frac{2}{665}(-241 + \sqrt{296401}) \approx 0.913 .$$

The ultraviolet behavior of the cosmological constant $\lambda(t)$, (4.128), crucially depends on the initial values of both the conformal coupling constant, $\nu^2(0)$, and the cosmological constant itself, $\lambda(0)$. In the region $\nu^2(0) > c_1 f^2(0)$ the solution (4.128) has a "zero-charge" pole at the same scale t_*, similarly to the conformal coupling constant $\nu^2(t)$, (4.125),

$$\lambda(t)\Big|_{t\to t_*} = \frac{3}{14} c_3 \frac{f_*^{2(1+2p)}}{f^{2p}(t) - f_*^{2p}} + O(1) . \qquad (4.130)$$

In the opposite case, $\nu^2(0) < c_1 f^2(0)$, the function $\lambda(t)$, (4.128), grows in the ultraviolet limit

$$\lambda(t)\Big|_{t\to\infty} = c_4 f^{-2q}(t) + O(f^2) . \qquad (4.131)$$

The sign of the constant c_4 in (4.131) depends on the initial value $\lambda(0)$, i.e., $c_4 > 0$ for $\lambda(0) > \lambda_2(0)$ and $c_4 < 0$ for $\lambda(0) > \lambda_2(0)$, where

$$\lambda_2(0) = -\Phi(0) \int\limits_0^\infty \mathrm{d}\tau A(\tau) \Phi^{-1}(\tau) \, . \tag{4.132}$$

In the special case $\lambda(0) = \lambda_2(0)$ the constant c_4 is equal to zero ($c_4 = 0$) and the solution (4.128) takes the form

$$\lambda(t) = \lambda_2(t) = -\Phi(t) \int\limits_t^\infty \mathrm{d}\tau A(\tau) \Phi^{-1}(\tau) \, . \tag{4.133}$$

The special solution (4.133) is asymptotically free in the ultraviolet limit

$$\lambda_2(t)\Big|_{t\to\infty} = c_5 f^2(t) + O\left(f^{2(1+p)}\right) , \tag{4.134}$$

where

$$c_5 = -\frac{5}{266} \cdot \frac{5 + c_2^2}{q+1} \approx -4.75 \, . \tag{4.135}$$

However, the special solution (4.133) is unstable because of the presence of growing mode (4.131). Besides, it exists only in the negative region $\lambda < 0$. In the positive region $\lambda > 0$ the cosmological constant is not asymptotically free, (4.131).

Our conclusions about the asymptotic behavior of the cosmological constant $\lambda(t)$ also differ essentially from the results of the papers [107, 108, 109, 111] where the asymptotic freedom for the cosmological constant in the region $\lambda > 0$ and $\nu^2 > 0$ for any initial values of $\lambda(0)$ was established.

Let us discuss the influence of arbitrary low-spin matter (except for spin-3/2 fields) interacting with the quadratic gravity (4.49) on the ultraviolet behavior of the theory. The system of renormalization group equations in presence of matter involves the equations (4.116)–(4.119) with the total β-function

$$\beta_{i,\mathrm{tot}} = \beta_i + \beta_{i,\mathrm{mat}} \, , \tag{4.136}$$

where $\beta_{i,\mathrm{mat}}$ is the contribution of matter fields in the gravitational divergences of the effective action, (4.71), and the equations for the masses and the matter coupling constants. The values of the first three coefficients at the terms quadratic in the curvature have the form [42, 129, 187]

$$\begin{aligned}
\beta_{1,\mathrm{mat}} &= -\frac{1}{360}\left(62N_1^{(0)} + 63N_1 + 11N_{1/2} + N_0\right) , \\
\beta_{2,\mathrm{mat}} &= \frac{1}{120}\left(12N_1^{(0)} + 13N_1 + 6N_{1/2} + N_0\right) , \\
\beta_{3,\mathrm{mat}} &= \frac{1}{72}\left(N_1 + (1-6\xi)^2 N_0\right) ,
\end{aligned} \tag{4.137}$$

where N_j is the number of the fields with spin j, $N_1^{(0)}$ is the number of massless vector fields, ξ is the coupling constant of scalar fields with the gravitational field. In the formula (4.137) the spinor fields are taken to be two-component. The coefficients (4.137) possess important general properties

$$\beta_{1,\mathrm{mat}} < 0 , \qquad \beta_{2,\mathrm{mat}} > 0 , \qquad \beta_{3,\mathrm{mat}} > 0 . \tag{4.138}$$

The gravitational β-function (4.72)–(4.74) obtained in previous sections have analogous properties for $f^2 > 0$ and $\nu^2 > 0$. Therefore, the total β-function, (4.136), also satisfy the conditions (4.138) for $f^2 > 0$ and $\nu^2 > 0$. The properties (4.138) are most important for the study of the ultraviolet asymptotics of the topological coupling constant $\epsilon(t)$, the Weyl coupling constant $f^2(t)$ and the conformal one $\nu^2(t)$.

The solution of the renormalization group equations for the topological and Weyl coupling constants in the presence of matter have the same form (4.120) with the substitution $\beta \to \beta_{\mathrm{tot}}$. Thus the presence of matter does not change qualitatively the ultraviolet asymptotics of these constants: the coupling $\epsilon(t)$ becomes negative and grows logarithmically and the Weyl coupling constant is asymptotically free at $f^2 > 0$.

The renormalization group equation for the conformal coupling constant $\nu^2(t)$ in the presence of the matter takes the form

$$\frac{d}{dt}\nu^2 = \frac{5}{3}f^4 + 5f^2\nu^2 + \frac{1}{12}\left(10 + N_1 + (1-6\xi)^2 N_0\right)\nu^4 . \tag{4.139}$$

Therefrom one can show that at $\nu^2 > 0$ the coupling constant $\nu^2(t)$ has a "zero-charge" singularity at a finite scale.

The other properties of the theory (in particular, the behavior of the conformal constant $\nu^2(t)$ at $\nu^2 < 0$) depend essentially on the particular form of the matter model. However, the strong conformal coupling, $\nu^2 \gg 1$, at $\nu^2 > 0$ leads to singularities in the cosmological constant as well as in all coupling constants of matter fields.

Thus, we conclude that the higher-derivative quantum gravity interacting with any low-spin matter necessarily goes out of the limits of weak conformal coupling at high energies in the case $\nu^2 > 0$. This conclusion is also opposite to the results of the papers [107, 108, 109, 111] where the asymptotic freedom of the higher-derivative quantum gravity in the region $\nu^2 > 0$ in the presence of rather arbitrary matter was established.

Let us also find the ultraviolet behavior of the non-essential Einstein coupling constant $k^2(t)$. The solution of the equation (4.119) has the form

$$k^2(t) = k^2(0) \exp\left\{ \int\limits_0^t d\tau \gamma(\tau) \right\} . \tag{4.140}$$

The explicit expression depends on the form of the function γ and, hence, on the gauge condition and the parametrization of the quantum field, (4.97). We will list the result for two cases: for the standard effective action in the minimal gauge and the standard parametrization (4.75) and for the Vilkovisky's effective action (4.114). In both cases the solution (4.140) has the form

$$k^2(t) = k^2(0)\frac{\Psi(t)}{\Psi(0)}\,, \tag{4.141}$$

where

$$\Psi(t) = \left|c_1 f^{2p}(t) - c_2 f_*^{2p}\right|^2 \left|f^{2p}(t) - f_*^{2p}\right|^s \left|f^2(t)\right|^{-r}\,,$$

$$s = \begin{cases} \frac{13}{5} \\ \\ \frac{3}{5} \end{cases}, \qquad r = \begin{cases} \frac{3}{665}(269 + \sqrt{296401}) \approx 3.67 \\ \\ \frac{2}{1995}(-437 + 2\sqrt{296401}) \approx 0.653 \end{cases}. \tag{4.142}$$

Here and below the upper values correspond to the Vilkovisky's effective action and the lower values correspond to the standard effective action in the minimal gauge and the standard parametrization.

Therefrom it is immediately seen that the Einstein coupling constant $k^2(t)$ grows in the ultraviolet limit $(t \to \infty)$

$$k^2(t)\Big|_{t\to\infty} = c_6 f^{-2r}(t) + O\left(f^{2(p-r)}\right)\,. \tag{4.143}$$

Let us note that the ultraviolet behavior of the dimensional cosmological constant, $\Lambda(t) = \lambda(t)/k^2(t)$, is essentially different in the case of the Vilkovisky's effective action and in the standard case

$$\Lambda(t)\Big|_{t\to\infty} = c_7 f^{2\alpha} + O\left(f^{2(\alpha+p)}\right)\,, \tag{4.144}$$

where

$$\alpha = r - q \approx \begin{cases} 2.76 \\ \\ -0.26 \end{cases}.$$

In the first case $\Lambda(t)$ rapidly approaches zero, like $\sim t^{-2.76}$, and in the second case it grows like $t^{0.26}$.

It is well known that the functional formulation of the quantum field theory assumes the Euclidean action to be positive definite [155, 193, 150]. Otherwise, (what happens, for example, in the conformal sector of the Einstein gravity), one must resort to the complexification of the configuration space to achieve the convergence of the functional integral [150, 131, 66].

The Euclidean action of the higher-derivative theory of gravity differs only by sign from the action (4.49) we are considering. It is positive definite in the case

$$\epsilon > 0\,, \tag{4.145}$$

$$\nu^2 < 0\,, \tag{4.146}$$

$$f^2 > 0, \qquad \lambda > -\frac{3}{4}\nu^2\,. \tag{4.147}$$

It is not necessary to impose the condition (4.145) if one restricts oneself to a fixed topology. However, if one includes in the functional integral of quantum gravity the topologically non-trivial metrics with large Euler characteristic, the violation of the condition (4.145) leads to the exponential growth of their weight and, therefore, to a foam-like structure of the space-time at micro-scales [150]. It is this situation that occurs in the ultraviolet limit, when $\epsilon(t) \to -\infty$, (4.120).

The condition (4.146) is usually held to be "non-physical" (see the bibliography). The point is, the conformal coupling constant ν^2 plays the role of the dimensionless square of the mass of the conformal mode on the flat background. In the case $\nu^2 < 0$ the conformal mode becomes tachyonic and leads to the instability of the flat space (i.e., oscillations of the static potential, unstable solutions etc. [209]).

As we showed above, the higher-derivative quantum gravity in the region $\nu^2 > 0$ has unsatisfactory "zero-charge" behavior in the conformal sector, (4.125), (4.130). In the region of strong conformal coupling ($\nu^2 \gg 1$) one cannot make definitive conclusions on the basis of the perturbative calculations. However, on the qualitative level it seems that the singularity in the coupling constants $\nu^2(t)$ and $\lambda(t)$ can be interpreted as a reconstruction of the ground state of the theory (phase transition), i.e., the conformal mode "freezes" and a conformal condensate is formed.

We find the arguments against the "non-physical" condition (4.146) to be not strong enough. First, the higher-derivative quantum gravity, strictly speaking, cannot be treated as a physical theory within the limits of perturbation theory because of the presence of the ghost states in the tensor sector that violate the unitarity of the theory (see the bibliography, in particular, [107, 108, 109, 111], [25]). This is not surprising in an asymptotically free theory (that always takes place in the tensor sector), since, generally speaking, the true physical asymptotic states have nothing to do with the excitations in the perturbation theory [215]. Second, the correspondence with the macroscopic gravitation is a rather fine problem that needs a special investigation of the low-energy limit of the higher-derivative quantum gravity. Third, the cosmological constant is always not asymptotically free. This means, presumably, that the expansion around the flat space in the high energy limit is not valid anymore. Hence, the solution of the unitarity problem based on this expansion by summing the radiation corrections and analyzing the position of the poles of the propagator in momentum representation is not valid too. In this case the flat background cannot present the ground state of the theory any longer.

From this standpoint, in high energy region the higher-derivative quantum gravity with positive definite Euclidean action, i.e., with an extra condition

$$\nu^2 < c_1 f^2 \approx -0.091 f^2, \tag{4.148}$$

seems to be more intriguing. Such theory has unique stable ground state that minimizes the functional of the classical Euclidean action. It is asymptotically free both in the tensor and conformal sectors. Besides, instead of the contradictory "zero-charge" behavior the cosmological constant just grows logarithmically at high energies.

Let us stress once more the main conclusion of the present section. Notwithstanding the fact that the higher-derivative quantum gravity is asymptotically free in the tensor sector of the theory with the natural condition $f^2 > 0$, that ensures the stability of the flat space under the tensor perturbations, the condition of the conformal stability of the flat background, $\nu^2 > 0$, is incompatible with the asymptotic freedom in the conformal sector. Thus, the flat background cannot present the ground state of the theory in the ultraviolet limit. The problem with the conformal mode does not appear in the conformally invariant models [118, 119]. Therefore, they are asymptotically free [107, 108, 109, 111, 110]; however, the appearance of the R^2-divergences at higher loops leads to their non-renormalizability [112].

4.5 Effective Potential

Up to now the background field (i.e., the space-time metric) was held to be arbitrary. To construct the S-matrix one needs the background fields to be the solutions of the classical equations of motion, (4.5), i.e., to lie on mass shell. It is obvious that the flat space does not solve the equations of motion (4.5) and (4.50) for non-vanishing cosmological constant. The most simple and maximally symmetric solution of the equations of motion (4.5) and (4.50) is the De Sitter space

$$R^{\mu}{}_{\nu\alpha\beta} = \frac{1}{12}(\delta^{\mu}_{\alpha} g_{\beta\nu} - \delta^{\mu}_{\beta} g_{\alpha\nu})R, \tag{4.149}$$

$$R_{\mu\nu} = \frac{1}{4} g_{\mu\nu} R, \qquad R = \text{const}\ ,$$

with the condition

$$R = 4\Lambda\ . \tag{4.150}$$

On the other hand, in quantum gravity De Sitter background, (4.149), plays the role of covariantly constant field strength in gauge theories, $\nabla_{\mu} R_{\alpha\beta\gamma\delta} = 0$. Therefore, the effective action on De Sitter background determines, in fact, the effective potential of the higher-derivative quantum gravity. Since in this case the background field is characterized only by one constant R, the effective potential is a usual function of one variable. The symmetry of De Sitter background makes it possible to calculate the one-loop effective potential exactly.

In the particular case of De Sitter background one can also check our result for the R^2-divergence of the one-loop effective action in general case, (4.71), i.e., the coefficient β_3, (4.73), that differs from the results of [107, 108,

109, 111] and radically changes the ultraviolet behavior of the theory in the conformal sector (see Sect. 4.4).

For the practical calculation of the effective potential we go to the Euclidean sector of the space-time. Let the space-time be a compact four-dimensional sphere S^4 with the volume

$$V = \int d^4x\, g^{1/2} = 24\left(\frac{4\pi}{R}\right)^2, \qquad (R > 0)\,, \tag{4.151}$$

and the Euler characteristic

$$\chi = \frac{1}{32\pi^2}\int d^4x\, g^{1/2} R^* R^* = 2\,. \tag{4.152}$$

In present section we will always use the Euclidean action that differs only by sign from the pseudo-Euclidean one, (4.49). All the formulas of the previous sections remain valid by changing the sign of the action S, the extremal $\varepsilon_i = S_{,i}$ and the effective action Γ.

On De Sitter background (4.149) the classical Euclidean action takes the form

$$S(R) = 24(4\pi)^2\left\{\frac{1}{6}\left(\epsilon - \frac{1}{\nu^2}\right) - \frac{1}{x} + 2\lambda\frac{1}{x^2}\right\}, \tag{4.153}$$

where $x \equiv Rk^2$ and $\lambda = \Lambda k^2$. For $\lambda > 0$ it has a minimum on the mass shell, $R = 4\Lambda$, $(x = 4\lambda)$, (4.150), that reads

$$S_{\text{on-shell}} = (4\pi)^2\left\{4\left(\epsilon - \frac{1}{\nu^2}\right) - \frac{3}{\lambda}\right\}. \tag{4.154}$$

Our aim is to obtain the effective value of the De Sitter curvature R from the full effective equations

$$\frac{\partial\Gamma(R)}{\partial R} = 0\,. \tag{4.155}$$

Several problems appear on this way: the dependence of the effective action and, therefore, the effective equations on the gauge condition and the parametrization of the quantum field, validity of the one-loop approximation etc. [113].

First of all, we make a change of field variables $h_{\mu\nu}$

$$h_{\mu\nu} = \bar{h}^{\perp}_{\mu\nu} + \frac{1}{4}g_{\mu\nu}\varphi + 2\nabla_{(\mu}\varepsilon_{\nu)}\,, \tag{4.156}$$

$$\varphi = h - \Box\sigma\,, \qquad h = g^{\mu\nu}h_{\mu\nu}\,, \tag{4.157}$$

$$\varepsilon_\mu = \varepsilon^{\perp}_{\mu} + \frac{1}{2}\nabla_\mu\sigma\,, \tag{4.158}$$

where the new variables, $\bar{h}^{\perp}_{\mu\nu}$ and $\varepsilon^{\perp}_{\mu}$, satisfy the differential constraints

$$\nabla^\mu \bar{h}^\perp_{\mu\nu} = 0\,, \qquad \bar{h}^\perp_{\mu\nu} g^{\mu\nu} = 0\,, \tag{4.159}$$

$$\nabla^\mu \varepsilon^\perp_\mu = 0\,. \tag{4.160}$$

In the following we will call the initial field variables $h_{\mu\nu}$, without any restrictions imposed on, the "unconstrained" fields and the fields $\bar{h}^\perp_{\mu\nu}$ and $\varepsilon^\perp_\mu$, which satisfy the differential conditions (4.159) and (4.160), the "constrained" ones.

When the unconstrained field is transformed under the gauge transformations with parameters ξ_μ, (4.43), (4.44),

$$\delta h_{\mu\nu} = 2\nabla_{(\mu}\xi_{\nu)}\,, \qquad \xi_\mu = \xi^\perp_\mu + \nabla_\mu \xi\,, \tag{4.161}$$

the constrained fields transform in the following way

$$\delta \bar{h}^\perp_{\mu\nu} = 0\,, \qquad \delta\varphi = 0\,, \tag{4.162}$$

$$\delta\varepsilon^\perp_\mu = \xi^\perp_\mu\,, \qquad \delta\sigma = 2\xi\,. \tag{4.163}$$

Therefore, the transverse traceless tensor field $\bar{h}^\perp_{\mu\nu}$ and the conformal field φ are the physical gauge-invariant components of the field $h_{\mu\nu}$, whereas $\varepsilon^\perp_\mu$ and σ are pure gauge non-physical ones.

The second variation of the Euclidean action on De Sitter background has the form

$$\begin{aligned} S_2(g+h) &\equiv \frac{1}{2} h^i S_{,ik} h^k = \int \mathrm{d}^4x\, g^{1/2} \\ &\times \left\{ \frac{1}{4f^2} \bar{h}^\perp \left[\Delta_2\left(m_2^2 + \frac{f^2+\nu^2}{3\nu^2}R\right) \Delta_2\left(\frac{R}{6}\right) + \frac{1}{2} m_2^2 (R-4\Lambda) \right] \bar{h}^\perp \right. \\ &- \frac{3}{32\nu^2} \left[\varphi \left(\Delta_0(m_0^2)\Delta_0\left(-\frac{R}{3}\right) + \frac{1}{3} m_0^2 (R-4\Lambda)\right) \varphi \right. \\ &\left. - \frac{2}{3} m_0^2 (R-4\Lambda)\varphi\Delta_0(0)\sigma - \frac{2}{3} m_0^2 (R-4\Lambda)\sigma\Delta_0(0)\Delta_0\left(-\frac{R}{2}\right)\sigma \right] \\ &\left. + \frac{1}{4k^2}(R-4\Lambda)\varepsilon^\perp \Delta_1\left(-\frac{R}{4}\right)\varepsilon^\perp \right\}\,, \end{aligned} \tag{4.164}$$

where

$$m_2^2 = \frac{f^2}{k^2}\,, \qquad m_0^2 = \frac{\nu^2}{k^2}\,,$$

and

$$\Delta_j(X) = -\Box + X \tag{4.165}$$

are the constrained differential operators acting on the constrained fields of spin $j = 0, 1, 2$.

When going to the mass shell, (4.150), the dependence of S_2, (4.164), on the non-physical fields $\varepsilon^{\perp}$ and σ disappears and (4.164) takes the form

$$S_2(g+h)\Big|_{\text{on-shell}}$$

$$= \int d^4x\, g^{1/2} \left\{ \frac{1}{4f^2} \bar{h}^{\perp} \Delta_2 \left(m_2^2 + \frac{4}{3} \frac{(f^2+\nu^2)}{\nu^2} \Lambda \right) \Delta_2 \left(\frac{2}{3} \Lambda \right) \bar{h}^{\perp} \right.$$

$$\left. - \frac{3}{32\nu^2} \varphi \Delta_0 (m_0^2) \Delta_0 \left(-\frac{4}{3} \Lambda \right) \varphi \right\} . \tag{4.166}$$

From here it follows, in particular, the gauge invariance of the second variation, (4.164), on the mass shell, (4.150),

$$\delta S_2(g+h)\Big|_{\text{on-shell}} = 0 . \tag{4.167}$$

The eigenvalues λ_n of the constrained Laplace operator $\Delta_j(0) = -\Box$ and their multiplicities d_n are [113]

$$\lambda_n = \rho^2 \bar{\lambda}_n , \qquad \bar{\lambda}_n = n^2 + 3n - j , \qquad \rho^2 = \frac{R}{12} ,$$

$$d_n = \frac{1}{6}(2j+1)(n+1-j)(n+2+j)(2n+3) , \qquad n = j, j+1, \dots . \tag{4.168}$$

Thus the condition of stability of the De Sitter background (4.149),

$$S_2(g+h)\Big|_{\text{on-shell}} > 0, \tag{4.169}$$

imposes the following restrictions (for $\lambda > 0$): in the tensor sector

$$f^2 > 0, \qquad -\frac{1}{f^2} - \frac{1}{3\nu^2} < \frac{1}{4\lambda} , \tag{4.170}$$

and in the conformal sector

$$\nu^2 < 0, \qquad \frac{1}{4\lambda} < -\frac{1}{3\nu^2} . \tag{4.171}$$

However, even in the case when these conditions are fulfilled there are still left five zero modes of the operator $\Delta_0 \left(-\frac{4}{3}\Lambda \right)$ in the conformal sector, (4.166). Along these conformal directions, φ_1, in the configuration space the Euclidean action does not grow

$$S_2(g+\varphi_1)\Big|_{\text{on-shell}} = 0. \tag{4.172}$$

This means that, in fact, the De Sitter background (4.149) does not give the absolute minimum of the positive definite Euclidean action. This can be verified by calculating the next terms in the expansion of $S(g+\varphi_1)$. However, in the one-loop approximation these terms do not matter.

To calculate the effective action one has to find the Jacobian of the change of variables (4.156)–(4.158). Using the simple equations

$$\int d^4x\, g^{1/2} h_{\mu\nu}^2 = \int d^4x\, g^{1/2} \left\{ \bar{h}^{\perp 2} + 2\varepsilon^\perp \Delta_1 \left(-\frac{R}{4}\right) \varepsilon^\perp \right.$$

$$\left. + \frac{3}{4}\sigma \Delta_0(0) \Delta_0 \left(-\frac{R}{3}\right) \sigma + \frac{1}{4} h^2 \right\}, \tag{4.173}$$

$$\int d^4x\, g^{1/2} \varepsilon_\mu^2 = \int d^4x\, g^{1/2} \left\{ \varepsilon^{\perp 2} + \frac{1}{4}\sigma \Delta_0(0) \sigma \right\}, \tag{4.173}$$

we obtain

$$dh_{\mu\nu} = d\bar{h}^\perp\, d\varepsilon^\perp\, d\sigma\, d\varphi\, (\det J_2)^{1/2} \,,$$
$$d\varepsilon_\mu = d\varepsilon^\perp\, d\sigma\, (\det J_1)^{1/2} \,, \tag{4.174}$$

where

$$J_2 = \Delta_1 \left(-\frac{R}{4}\right) \otimes \Delta_0 \left(-\frac{R}{3}\right) \otimes \Delta_0(0) \,,$$
$$J_1 = \Delta_0(0) \,.$$

Let us calculate the effective action (4.12) in De Witt gauge (4.56). The ghost operator F, (4.16), in this gauge equals the operator N, (4.48). On De Sitter background (4.149) it has the form

$$F_{\mu\nu} = N_{\mu\nu} = 2g^{1/2} \left\{ g_{\mu\nu} \left(-\Box - \frac{R}{4} \right) + \frac{1}{2}(\kappa - 1)\nabla_\mu \nabla_\nu \right\} \delta(x,y) \,. \tag{4.175}$$

The operator of "averaging over the gauges" H, (4.58), and the gauge fixing term have the form

$$H^{\mu\nu} = \frac{1}{4\alpha^2} g^{-1/2} \left\{ g^{\mu\nu} \left(-\Box + \frac{R+P}{4} \right) + \beta \nabla^\mu \nabla^\nu \right\} \delta(x,y) \,, \tag{4.176}$$

$$S_{\text{gauge}} \equiv \frac{1}{2} \chi_\mu H^{\mu\nu} \chi_\nu$$

$$= \int d^4x\, g^{1/2} \frac{1}{2\alpha^2} \left\{ \varepsilon^\perp \Delta_1 \left(\frac{R+P}{4} \right) \Delta_1^2 \left(-\frac{R}{4} \right) \varepsilon^\perp \right.$$

$$+\frac{1}{16}(1-\beta)\Bigg[\kappa^2\varphi\Delta_0(P')\Delta_0(0)\varphi$$

$$-2\kappa(\kappa-3)\varphi\Delta_0(P')\Delta_0\left(\frac{R}{\kappa-3}\right)\Delta_0(0)\sigma$$

$$+(\kappa-3)^2\sigma\Delta_0(P')\Delta_0^2\left(\frac{R}{\kappa-3}\right)\Delta_0(0)\sigma\Bigg]\Bigg\}\,, \tag{4.177}$$

where

$$P = g_{\mu\nu}P^{\mu\nu} = (p_1+4p_2)R + 4p_3\frac{1}{k^2}\,, \qquad P' \equiv \frac{P}{1-\beta}\,.$$

Using the equation

$$\int \mathrm{d}^4x\, g^{1/2}\varepsilon_\mu \left\{g^{\mu\nu}\left(-\Box + X\right) + \gamma\nabla^\mu\nabla^\nu\right\}\varepsilon_\nu$$

$$= \int \mathrm{d}^4x\, g^{1/2}\left\{\varepsilon^\perp\Delta_1(X)\varepsilon^\perp + \frac{1-\gamma}{4}\sigma\Delta_0\left(\frac{X-\frac{R}{4}}{1-\gamma}\right)\Delta_0(0)\sigma\right\} \tag{4.178}$$

we find the determinants of the ghost operators (4.175) and (4.176)

$$\det F = \det\Delta_1\left(-\frac{R}{4}\right)\det\Delta_0\left(\frac{R}{\kappa-3}\right)\,,$$

$$\det H = \det\Delta_1\left(\frac{R+P}{4}\right)\det\Delta_0\left(P'\right)\,. \tag{4.179}$$

For the operators F and H to be positive definite we assume $\beta < 1$ and $\kappa < 3$.

Thus using the determinants of the ghost operators, (4.179), and the Jacobian of the change of variables, (4.174), and integrating $\exp(-S_2 - S_{\text{gauge}})$ we obtain the one-loop effective action off mass shell in De Witt gauge (4.56) with arbitrary gauge parameters κ, α, β and P

$$\Gamma_{(1)} = I_{(2)} + I_{(1)} + I_{(0)}\,, \tag{4.180}$$

where

$$I_{(2)} = \frac{1}{2}\log\det\left[\Delta_2\left(\frac{R}{6}\right)\Delta_2\left(m_2^2 + \frac{f^2+\nu^2}{3\nu^2}R\right) + \frac{1}{2}m_2^2(R-4\Lambda)\right]\,, \tag{4.181}$$

$$I_{(1)} = \frac{1}{2}\log\frac{\det\left[\Delta_1\left(-\frac{R}{4}\right)\Delta_1\left(\frac{R+P}{4}\right) + \frac{1}{2}\frac{\alpha^2}{k^2}(R-4\Lambda)\right]}{\left(\det\Delta_1\left(-\frac{R}{4}\right)\right)^2\det\Delta_1\left(\frac{R+P}{4}\right)}\,, \tag{4.182}$$

$$I_{(0)} = \frac{1}{2} \log \frac{\det \left[\Delta_0^2 \left(\frac{R}{\kappa - 3} \right) \Delta_0 \left(P' \right) \Delta_0 \left(m_0^2 \right) + \mathcal{D} \left(\Lambda - \frac{R}{4} \right) + \mathcal{C} \left(\Lambda - \frac{R}{4} \right)^2 \right]}{\left(\det \Delta_0 \left(\frac{R}{\kappa - 3} \right) \right)^2 \det \Delta_0 \left(P' \right)} , \tag{4.183}$$

$$\mathcal{D} = 4 m_0^2 \frac{\kappa^2 - 3}{(\kappa - 3)^2} \Delta_0 \left(\frac{R}{\kappa^2 - 3} \right) \Delta_0 \left(P' \right)$$

$$- \frac{8\alpha^2}{k^2 (1 - \beta)(\kappa - 3)^2} \Delta_0 \left(m_0^2 \right) \Delta_0 \left(- \frac{R}{2} \right) , \tag{4.184}$$

$$\mathcal{C} = 16 \frac{\alpha^2 \nu^2}{k^4 (1 - \beta)(\kappa - 3)^2} .$$

The quantity $I_{(2)}$ describes the contribution of two tensor fields, $I_{(1)}$ gives the contribution of the vector ghost and $I_{(0)}$ is the contribution of the scalar conformal field off mass shell. The contribution of the tensor fields $I_{(2)}$, (4.181), does not depend on the gauge, the contribution of the vector ghost $I_{(1)}$, (4.182), depends on the parameters α and P and the contribution of the scalar field $I_{(0)}$, (4.183), (4.184), depends on all gauge parameters κ, α, β and P.

The expressions for $I_{(1)}$ and $I_{(0)}$, (4.182), (4.183), are simplified in some particular gauges

$$I_{(1)} \Big|_{\alpha=0} = - \frac{1}{2} \log \det \Delta_1 \left(- \frac{R}{4} \right) , \tag{4.185}$$

$$I_{(0)} \Big|_{\substack{\alpha=0 \\ \kappa=0}} = I_{(0)} \Big|_{\substack{\beta=-\infty \\ \kappa=0}} = \frac{1}{2} \log \frac{\det \left[\Delta_0 \left(- \frac{R}{3} \right) \Delta_0 \left(m_0^2 \right) + \frac{1}{3} m_0^2 (R - 4\Lambda) \right]}{\det \Delta_0 \left(- \frac{R}{3} \right)} , \tag{4.186}$$

$$I_{(0)} \Big|_{\substack{\alpha=0 \\ \kappa=1}} = I_{(0)} \Big|_{\substack{\beta=-\infty \\ \kappa=1}} = \frac{1}{2} \log \frac{\det \left[\Delta_0 \left(- \frac{R}{2} \right) \Delta_0 \left(m_0^2 \right) + \frac{1}{2} m_0^2 (R - 4\Lambda) \right]}{\det \Delta_0 \left(- \frac{R}{2} \right)} , \tag{4.187}$$

$$I_{(0)} \Big|_{\kappa=-\infty} = \frac{1}{2} \log \frac{\det \left[\Delta_0(0) \Delta_0 \left(m_0^2 \right) - m_0^2 (R - 4\Lambda) \right]}{\det \Delta_0(0)} , \tag{4.188}$$

Let us also calculate the gauge-independent and reparametrization-invariant Vilkovisky's effective action, (4.21), on De Sitter background in the orthogonal gauge, (4.36), (4.38). In the one-loop approximation, (4.40), it differs from the standard effective action in De Witt gauge, (4.14), (4.56), (4.58), with $\alpha = 0$ only by an extra term in the operator $\tilde{\Delta}$, (4.41), due to the Christoffel connection of the configuration space, (4.27), (4.47). Therefore, the Vilkovisky's effective action, (4.38), (4.40), can be obtained from the standard one, (4.12), (4.14), in De Witt gauge, (4.56), (4.58), for $\alpha = 0$ by substituting the operator $\mathcal{D}_i\mathcal{D}_k S$, (4.41), for the operator $S_{,ik}$, i.e., by the replacing the quadratic part of the action $S_2(g+h)$, (4.164), by

$$\tilde{S}_2(g+h) = S_2(g+h) - \frac{1}{2}h^i \left\{ {j \atop ik} \right\} \varepsilon_j h^k ,$$

$$h^i \left\{ {j \atop ik} \right\} \varepsilon_j h^k = -\frac{1}{4}(\kappa^{-1}-1)\frac{1}{k^2}(R-4\Lambda)\int \mathrm{d}^4x\, g^{1/2}\left(h^2_{\mu\nu} - \frac{1}{4}h^4\right)$$

$$= -\frac{1}{4}(\kappa^{-1}-1)\frac{1}{k^2}(R-4\Lambda)\int \mathrm{d}^4x\, g^{1/2}\Bigg\{\bar{h}^{\perp\, 2}$$

$$+2\varepsilon^\perp \Delta_1\left(-\frac{R}{4}\right)\varepsilon^\perp + \frac{3}{4}\sigma\Delta_0(0)\Delta_0\left(-\frac{R}{3}\right)\sigma\Bigg\} , \quad (4.189)$$

where κ is the parameter of the configuration space metric that is given, according to the paper [223]], by the formula (4.108). Let us note that in the Einstein gravity one obtains for the parameter κ the value $\kappa = 1$ [223, 224]. Therefore, the additional contribution of the connection (4.189) vanishes and the Vilkovisky's effective action on De Sitter background coincides with the standard one computed in De Witt gauge, (4.56), (4.58), with $\alpha = 0$ and $\kappa = 1$ (i.e., in the harmonic De Donder–Fock–Landau gauge (4.60)).

Taking into account the Jacobian of the change of variables (4.174) and the ghosts determinant (4.179) the functional measure in the constrained variables takes the form

$$dh_{\mu\nu}\delta(R_{i\mu}h^i)\det N = d\bar{h}^\perp\, d\varepsilon^\perp\, d\varphi\, d\sigma\, \delta(\varepsilon^\perp)$$

$$\times\, \delta\left[\kappa\varphi - (\kappa-3)\Delta_0\left(\frac{R}{\kappa-3}\right)\sigma\right]\det\Delta_0\left(\frac{R}{\kappa-3}\right)$$

$$\times\left[\det\Delta_1\left(-\frac{R}{4}\right)\det\Delta_0\left(-\frac{R}{3}\right)\right]^{1/2} . \quad (4.190)$$

Integrating $\exp(-\tilde{S}_2)$ we obtain the one-loop Vilkovisky's effective action

$$\tilde{\Gamma}_{(1)} = \tilde{I}_{(2)} + \tilde{I}_{(1)} + \tilde{I}_{(0)} , \quad (4.191)$$

where

$$\tilde{I}_{(2)} = \frac{1}{2}\log\det\left[\Delta_2\left(\frac{R}{6}\right)\Delta_2\left(m_2^2 + \frac{f^2+\nu^2}{3\nu^2}R\right) + \frac{1}{2\kappa}m_2^2(R-4\Lambda)\right] , \quad (4.192)$$

$$\tilde{I}_{(1)} = -\frac{1}{2}\log\det\Delta_1\left(-\frac{R}{4}\right) , \quad (4.193)$$

$$\tilde{I}_{(0)} = \frac{1}{2}\log\frac{\det\left[\Delta_0\left(\frac{R}{\kappa-3}\right)\Delta_0\left(m_0^2\right) - \frac{1}{\kappa-3}m_0^2(R-4\Lambda)\right]}{\det\Delta_0\left(\frac{R}{\kappa-3}\right)} . \tag{4.194}$$

The equations (4.191)–(4.194) for $\kappa = 1$ do coincide indeed with the standard effective action in the gauge $\alpha = 0$, $\kappa = 1$, (4.180), (4.181), (4.185), (4.187). However, to obtain the Vilkovisky's effective action in our case one has to put in (4.191)–(4.194) $\kappa = 3f^2/(f^2 + 2\nu^2)$, (4.108).

On the mass shell (4.150) the dependence on the gauge disappears and we have

$$I_{(2)}^{\text{on-shell}} = \frac{1}{2}\log\det\Delta_2\left(\frac{2}{3}\Lambda\right) + \frac{1}{2}\log\det\Delta_2\left(m_2^2 + \frac{4}{3}\cdot\frac{(f^2+\nu^2)}{\nu^2}\Lambda\right) , \tag{4.195}$$

$$I_{(1)}^{\text{on-shell}} = -\frac{1}{2}\log\det\Delta_1\left(-\Lambda\right) , \tag{4.196}$$

$$I_{(0)}^{\text{on-shell}} = \frac{1}{2}\log\det\Delta_0(m_0^2) . \tag{4.197}$$

From here one sees immediately the spectrum of the physical excitations of the theory: one massive tensor field of spin 2 (5 degrees of freedom), one massive scalar field (1 degree of freedom) and the Einstein graviton, i.e., the massless tensor field of spin 2 ($5 - 3 = 2$ degrees of freedom). Altogether the higher-derivative quantum gravity (4.49) has $5 + 2 + 1 = 8$ degrees of freedom.

To calculate the functional determinants of the differential operators we will use the technique of the generalized ζ-function [129, 113, 150, 149, 89, 131, 66]. Let us define the ζ-function by the functional trace of the complex power of the differential operator of order $2k$, $\Delta^{(k)}$,

$$\zeta_j\left(p; \Delta^{(k)}/\mu^{2k}\right) \equiv \operatorname{tr}\left(\Delta^{(k)}/\mu^{2k}\right)^{-p} , \tag{4.198}$$

where

$$\Delta^{(k)} = P_k(-\Box) , \tag{4.199}$$

$P_k(x)$ is a polynomial of order k, μ is a dimensional mass parameter and j denotes the spin of the field, the operator $\Delta^{(k)}$ is acting on.

For $\operatorname{Re} p > 2/k$ the ζ-function is determined by the convergent series over the eigenvalues (4.168)

$$\zeta_j\left(p; \Delta^{(k)}/\mu^{2k}\right) = \sum_n d_n\left(P_k(\lambda_n)/\mu^{2k}\right)^{-p} , \tag{4.200}$$

where the summation runs over all modes of the Laplace operator, (4.168), with positive multiplicities, $d_n > 0$, including the negative and zero modes of the operator $\Delta^{(k)}$. The zero modes give an infinite constant that should be simply subtracted, whereas the negative modes lead to an imaginary part indicating to the instability [113]. For $\mathrm{Re}\, p \le 2/k$ the analytical continuation of (4.200) defines a meromorphic function with poles on the real axis. It is important, that the ζ-function is analytic at the point $p = 0$. Therefore, one can define the finite values of the total number of modes of the operator $\Delta^{(k)}$ (taking each mode k times) and its functional determinant

$$k \,\mathrm{tr}\, 1 = B(\Delta^{(k)}) \,, \tag{4.201}$$

$$\log\det\left(\Delta^{(k)}/\mu^{2k}\right) = -\zeta_j'(0) \,, \tag{4.202}$$

where

$$B(\Delta^{(k)}) = k\zeta_j(0) \,,$$

$$\zeta_j'(p) = \frac{d}{dp}\zeta_j(p) \,. \tag{4.203}$$

Under the change of the scale parameter μ the functional determinant behaves as follows

$$\zeta_j'\left(0; \Delta^{(k)}/\mu^{2k}\right) = -B(\Delta^{(k)}) \log\frac{\rho^2}{\mu^2} + \zeta_j'\left(0; \Delta^{(k)}/\rho^{2k}\right) \,. \tag{4.204}$$

Using the spectrum of the Laplace operator (4.168) we rewrite (4.200) in the form

$$\zeta_j\left(p; \Delta^{(k)}/\rho^{2k}\right) = \frac{2j+1}{3} \sum_{\nu \ge j+\frac{3}{2}\,,\ \Delta\nu = 1} \nu(\nu^2 - l^2)$$

$$\times \left[P_k\left(\rho^2\left(\nu^2 - \frac{9}{4} - j\right)\right)\right]^{-p} \,, \tag{4.205}$$

where

$$\rho^2 = \frac{R}{12} \,, \qquad l = j + \frac{1}{2} \,.$$

The sum (4.205) can be calculated for $\mathrm{Re}\, p > \frac{2}{k}$ by means of the Abel–Plan summation formula [98]

$$\sum_{\nu \ge \frac{1}{2}} f(\nu) = \sum_{\frac{1}{2} \le \nu \le k - \frac{1}{2}} f(\nu) + \int\limits_{k+\varepsilon}^{\infty} \mathrm{d}t\, f(t)$$

$$+ \int\limits_0^{\infty} \frac{\mathrm{d}t}{\mathrm{e}^{2\pi t} + 1} \left[\mathrm{i} f(k + \varepsilon - \mathrm{i}t) - \mathrm{i} f(k + \varepsilon + \mathrm{i}t)\right] , \tag{4.206}$$

where the integer k should be chosen in such a way that for $\mathrm{Re}\,\nu > k$ the function $f(\nu)$ is analytic, i.e., it does not have any poles. The infinitesimal parameter $\varepsilon > 0$ shows the way how to get around the poles (if any) at $\operatorname{Re}\nu = k$. The formula (4.206) is valid for the functions $f(\nu)$ that fall off sufficiently rapidly at the infinity:

$$f(\nu)\Big|_{|\nu|\to\infty} \sim |\nu|^{-q}, \qquad \operatorname{Re} q > 1. \tag{4.207}$$

When applying the formula (4.206) to (4.205) the second integral in (4.206) gives an analytic function of the variable p. All the poles of the ζ-function are contained in the first integral. By using the analytical continuation and integrating by parts one can calculate both $\zeta(0)$ and $\zeta'(0)$.

As a result we obtain for the operator of second order,

$$\Delta_j(X) = -\Box + X, \tag{4.208}$$

and for the operator of forth order,

$$\Delta_j^{(2)}(X,Y) = \Box^2 - 2X\Box + Y, \tag{4.209}$$

the finite values of the total number of modes (4.201) and the determinant (4.202)

$$B(\Delta_j(X)) = \frac{2j+1}{12}\left\{(b^2+l^2)^2 - \frac{2}{3}l^2 + \frac{1}{30}\right\}, \tag{4.210}$$

$$B\left(\Delta_j^{(2)}(X,Y)\right) = 2\cdot\frac{2j+1}{12}\left\{(b^2+l^2)^2 - \frac{2}{3}l^2 + \frac{1}{30} - a^2\right\}, \tag{4.211}$$

$$\zeta_j'(0;\Delta_j(X)/\rho^2) = \frac{2j+1}{3}F_j^{(0)}(\bar{X}), \tag{4.212}$$

$$\zeta_j'(0;\Delta_j^{(2)}(X,Y)/\rho^4) = \frac{2j+1}{3}F_j^{(2)}(\bar{X},\bar{Y}), \tag{4.213}$$

where

$$b^2 = \bar{X} - j - \frac{9}{4}, \qquad \bar{X} = \frac{X}{\rho^2},$$

$$a^2 = \bar{Y} - \bar{X}^2, \qquad \bar{Y} = \frac{Y}{\rho^4},$$

and

$$F_j^{(0)}(\bar{X}) = -\frac{1}{4}b^2(b^2+2l^2)\log b^2 + \frac{1}{2}l^2b^2 + \frac{3}{8}b^4$$

$$+2\int_0^\infty \frac{\mathrm{d}t\, t}{e^{2\pi t}+1}(t^2+l^2)\log|b^2-t^2|$$

$$+\sum_{\frac{1}{2}\le\nu\le j-\frac{1}{2}} \nu(\nu^2-l^2)\log(\nu^2+b^2), \tag{4.214}$$

$$F_j^{(2)}(\bar{X},\bar{Y}) = \frac{1}{4}(a^2 - b^4 - 2l^2b^2)\log(b^4 + a^2)$$

$$-a(b^2 + l^2)\left[\arctan\left(\frac{b^2}{a}\right) - \frac{\pi}{2}\right] - \frac{1}{4}a^2 + \frac{3}{4}b^4 + l^2b^2$$

$$+2\int_0^\infty \frac{\mathrm{d}t\, t}{\mathrm{e}^{2\pi t} + 1}(t^2 + l^2)\log\left[(b^2 - t^2)^2 + a^2\right]$$

$$+ \sum_{\frac{1}{2}\le\nu\le j-\frac{1}{2}} \nu(\nu^2 - l^2)\log\left[(\nu^2 + b^2)^2 + a^2\right] . \qquad (4.215)$$

The introduced functions, (4.214) and (4.215), are related by the equation

$$F_j^{(0)}(\bar{X}) + F_j^{(0)}(\bar{Y}) = F_j^{(2)}\left(\frac{\bar{X}+\bar{Y}}{2}; \bar{X}\bar{Y}\right) . \qquad (4.216)$$

In complete analogy one can obtain the functional determinants of the operators of higher orders and even non-local, i.e., integro-differential, operators.

Using the technique of the generalized ζ-function and separating the dependence on the renormalization parameter μ we get the one-loop effective action, (4.180)–(4.184),

$$\Gamma_{(1)} = \frac{1}{2}B_{\mathrm{tot}}\log\frac{R}{12\mu^2} + \Gamma_{(1)\ \mathrm{ren}} , \qquad (4.217)$$

where

$$\Gamma_{(1)\ \mathrm{ren}} = \Gamma_{(1)}\Big|_{\mu^2=\rho^2=\frac{R}{12}} . \qquad (4.218)$$

For the study of the effective action one should calculate, first of all, the coefficient B_{tot}. To calculate the contributions of the tensor, (4.181), and vector, (4.182), fields in B_{tot} it suffices to use the formulas (4.210) and (4.211) for the operators of the second and the forth order. Although the contribution of the scalar field in arbitrary gauge, (4.183), (4.184), contains an operator of eighth order, it is not needed to calculate the coefficient B for the operator of the eighth order. Noting that on mass shell, (4.150), the contribution of the scalar field (4.197) contains only a second order operator, one can expand the contribution of the scalar field off mass shell in the extremal, i.e., in $(R - 4\Lambda)$, limiting oneself only to linear terms.

One should note, that the differential change of the variables (4.156)–(4.158) brings some new zero modes that were not present in the non-constrained operators. Therefore, when calculating the total number of modes (i.e., the coefficient B_{tot}) one should subtract the number of zero modes of the Jacobian of the change of variables (4.174):

$$B_{\rm tot} = \sum_i B(\Delta_i) - \mathcal{N}(J) \,. \tag{4.219}$$

Using the number of zero modes of the operators entering the Jacobians, (4.174),

$$\mathcal{N}(\Delta_0(0)) = 1\,, \qquad \mathcal{N}\left(\Delta_0\left(-\frac{R}{3}\right)\right) = 5\,, \qquad \mathcal{N}\left(\Delta_1\left(-\frac{R}{4}\right)\right) = 10\,, \tag{4.220}$$

we obtain

$$\mathcal{N}(J) = 2\times 15 - 1 = 29\,. \tag{4.221}$$

Thus we obtain the coefficient $B_{\rm tot}$

$$\begin{aligned} B_{\rm tot} &= \frac{20}{3}\frac{f^4}{\nu^4} + 20\frac{f^2}{\nu^2} - \frac{634}{45} + 24\gamma\frac{1}{x} \\ &+16\left[\frac{3}{4}(\nu^4+5f^4) + \lambda\left(10\frac{f^4}{\nu^2} + 15f^2 - \nu^2 - 6\gamma\right)\right]\frac{1}{x^2}\,, \end{aligned} \tag{4.222}$$

where $x = Rk^2$ and the coefficient γ is given by the formula (4.97).

The Vilkovisky's effective action (4.191) has the same form, (4.217), with the coefficient $\tilde{B}_{\rm tot}$ of the form (4.222) but with the change $\gamma \to \tilde{\gamma}$, where $\tilde{\gamma}$ is given by the formula (4.114).

On the other hand, the coefficient $B_{\rm tot}$ can be obtained from the general expression for the divergences of the effective action (4.71) on De Sitter background (4.149),

$$B_{\rm tot} = 4\beta_1 + 24\beta_3 + 24\gamma\frac{1}{x} + 16\left(\frac{3}{2}\beta_4 - 6\gamma\lambda\right)\frac{1}{x^2}\,, \tag{4.223}$$

where the coefficients β_1, β_3 and β_4 are given by the formulas (4.72)–(4.74), and the coefficient γ is given by the formula (4.97) for standard effective action in arbitrary gauge and by the expression (4.114) for the Vilkovisky's effective action.

Comparing the expressions (4.222) and (4.223) we convince ourselves that our result for the coefficient β_3, (4.73), that differs from the results of other authors [107, 108, 109, 111] (see Sect. 4.2), and our results for the divergences of the effective action in arbitrary gauge, (4.97), and for the divergences of the Vilkovisky's effective action, (4.114), are correct.

On mass shell (4.150) the coefficient (4.222) does not depend on the gauge and we have a single-valued expression

$$\begin{aligned} B_{\rm tot}^{\rm on-shell} &= \frac{20}{3}\frac{f^4}{\nu^4} + 20\frac{f^2}{\nu^2} - \frac{634}{45} \\ &+ \left(10\frac{f^4}{\nu^2} + 15f^2 - \nu^2\right)\frac{1}{\lambda} + \frac{3}{4}(\nu^4+5f^4)\frac{1}{\lambda^2}\,. \end{aligned} \tag{4.224}$$

Let us also calculate the finite part of the effective action (4.218). Since it depends essentially on the gauge, (4.180)–(4.184), we limit ourselves to the case of the Vilkovisky's effective action, (4.191)–(4.194). Using the results (4.202) and (4.212)–(4.215) we obtain

$$\tilde{\Gamma}_{(1)\text{ren}} = -\frac{1}{6}\left\{5F_2^{(2)}(Z_1, Z_2) - 3F_1^{(0)}(-3)\right.$$

$$\left. + F_0^{(2)}(Z_3, Z_4) - F_0^{(0)}\left(-2\frac{f^2}{\nu^2} - 4\right)\right\} , \qquad (4.225)$$

where

$$Z_1 = 6f^2\frac{1}{x} + 2\frac{f^2}{\nu^2} + 3 ,$$

$$Z_2 = -96(f^2 + 2\nu^2)\lambda\frac{1}{x^2} + 48(f^2 + \nu^2)\frac{1}{x} + 8\frac{f^2}{\nu^2} + 8 , \qquad (4.226)$$

$$Z_3 = 6\nu^2\frac{1}{x} - \frac{f^2}{\nu^2} - 2 ,$$

$$Z_4 = -96(f^2 + 2\nu^2)\lambda\frac{1}{x^2} ,$$

$F_j^{(k)}$ are the functions introduced above, (4.214), (4.215), and $x = Rk^2$.

On mass shell (4.150) the effective action does not depend on the gauge and has the form, (4.13), (4.154), (4.195)-(4.197), (4.217),

$$\Gamma_{\text{on-shell}} = (4\pi)^2\left\{4\left(\epsilon - \frac{1}{\nu^2}\right) - \frac{3}{\lambda}\right\}$$

$$+\hbar\left\{\frac{1}{2}B_{\text{tot}}^{\text{on-shell}}\log\frac{\lambda}{3\mu^2k^2} + \Gamma_{(1)\text{ren}}^{\text{on-shell}}\right\} + O(\hbar^2) , \qquad (4.227)$$

where

$$\Gamma_{(1)\text{ren}}^{\text{on-shell}} = -\frac{1}{6}\left\{5F_2^{(0)}\left(3\frac{f^2}{\lambda} + 4\frac{f^2}{\nu^2} + 4\right)\right.$$

$$\left. + 5F_2^{(0)}(2) - 3F_1^{(0)}(-3) + F_0^{(0)}\left(3\frac{\nu^2}{\lambda}\right)\right\} . \qquad (4.228)$$

The expression (4.227) gives the vacuum action on De Sitter background with quantum corrections. It is real, since the operators in (4.195)–(4.197) do not have any negative modes provided the conditions (4.170) and (4.171) are fulfilled. Although the operator $\Delta_0(m_0^2)$ has one negative mode, $\varphi = \text{const}$, subject to the condition (4.171), it is non-physical, since it is just the zero mode of the Jacobian of the change of variables (4.174). All other modes of the operator $\Delta_0(m_0^2)$ are positive subject to the condition (4.171) in spite of the fact that $m_0^2 < 0$. Depending on the value of De Sitter curvature R off mass shell there can appear negative modes leading to an imaginary part of the effective action.

Differentiating the effective action, (4.13), (4.153), (4.217), (4.222), (4.225), we obtain the effective equation for the background field, i.e., the curvature of De Sitter space,

$$\frac{1}{\hbar k^2}\frac{\partial\Gamma}{\partial R} = \frac{1}{\hbar}24(4\pi)^2\cdot\frac{x-4\lambda}{x^3} + \frac{1}{2x}B_{\rm tot}(x)$$

$$+\frac{1}{2}B'_{\rm tot}(x)\log\frac{x}{12\mu^2k^2} + \Gamma'_{(1)\rm ren}(x) + O(\hbar) = 0\,, \quad (4.229)$$

where

$$B'_{\rm tot}(x) = \frac{\partial B_{\rm tot}}{\partial x}$$

$$= -\frac{32}{x^3}\left[\frac{3}{4}(\nu^4+5f^4)+\lambda\left(10\frac{f^4}{\nu^2}+15f^2-\nu^2-6\gamma\right)\right]$$

$$-24\gamma\frac{1}{x^2}\,, \quad (4.230)$$

$$\Gamma'_{(1)\rm ren}(x) = \frac{\partial\Gamma_{(1)\rm ren}}{\partial x}\,, \qquad x = Rk^2\,. \quad (4.231)$$

The perturbative solution of the effective equation (4.229) has the form

$$Rk^2 = 4\lambda - \hbar\frac{\lambda^2}{3(4\pi)^2}\left\{B_{\rm tot}^{\rm on-shell} + 4\lambda B'_{\rm tot}(4\lambda)\log\frac{\lambda}{3\mu^2k^2}\right.$$

$$\left. +8\lambda\Gamma'_{(1)\rm ren}(4\lambda)\right\}+O(\hbar^2)\,. \quad (4.232)$$

It gives the corrected value of the curvature of De Sitter space with regard to the quantum effects.

perturbation theory near this solution is applicable for $\lambda \neq 0$ in the region $f^2 \sim \nu^2 \sim \lambda \ll 1$. For $\lambda \sim 1$, i.e., when Λ is of Planck mass order $1/k^2$, the contributions of higher loops are essential and the perturbation theory is not adequate anymore.

Apart from the perturbative solution (4.232) the equation (4.229) can also have non-perturbative ones. In the special case $\lambda = 0$ the non-perturbative solution $R \neq 0$ means the spontaneous creation of De Sitter space from the flat space due to quantum gravitational fluctuations. Therefore, it seems quite possible that De Sitter space, needed in the inflational cosmological scenarios of the evolution of the Universe [171]], has quantum-gravitational origin [139]. However, almost any non-perturbative solution has the order $Rk^2 \sim 1$ and, therefore, is inapplicable in the one-loop approximation.

Conclusion

Let us summarize shortly the main results.

1. The methods for the covariant expansions of arbitrary fields in a curved space with arbitrary connection in generalized Taylor series and the Fourier integral in most general form are formulated.
2. A manifestly covariant technique for the calculation of De Witt coefficients based on the method of covariant expansions is elaborated. The corresponding diagrammatic formulation of this technique is given.
3. The De Witt coefficients a_3 and a_4 at coinciding points are calculated.
4. The renormalized one-loop effective action for the massive scalar, spinor and vector fields in an background gravitational field up to the terms of order $1/m^4$ is calculated.
5. Covariant methods for studying the non-local structure of the effective action are developed.
6. The terms of first order in background fields in De Witt coefficients are calculated. The summation of these terms is carried out and a non-local covariant expression for the Green function at coinciding points up to terms of second order in background fields is obtained. It is shown that in the conformally invariant case the Green function at coinciding points is finite in the first order in the background fields.
7. The terms of second order in background fields in De Witt coefficients are calculated. The summation of these terms is carried out and a manifestly covariant non-local expression for the one-loop effective action up to the terms of third order in background fields is obtained. All formfactors, their asymptotics and imaginary parts (for standard definition of the asymptotic regions, ground states and causal boundary conditions) are calculated. A finite effective action in the conformally invariant case of massless scalar field in two-dimensional space is obtained.
8. The De Witt coefficients for the case of scalar field in De Sitter space are calculated. It is shown that the corresponding Schwinger–De Witt series diverges. The Borel summation of the Schwinger–De Witt expansion is carried out and an explicit non-analytic expression in the background fields for the one-loop effective action is obtained.

9. The off-shell one-loop divergences of the effective action in arbitrary covariant gauge as well as those of the Vilkovisky's effective action in higher-derivative quantum gravity are calculated.
10. The ultraviolet asymptotics of the coupling constants of the higher-derivative quantum gravity are found. It is shown that in the "physical" region of the coupling constants, that is characterized by the absence of the tachyons on the flat background, the conformal sector has "zero-charge" behavior. Therefore, the higher-derivative quantum gravity at higher energies goes beyond the limits of weak conformal coupling. This conclusion does not depend on the presence of the matter fields of low spins. In other words, the condition of the conformal stability of the flat background, which is held usually as "physical", is incompatible with the asymptotic freedom in the conformal sector. Therefore, the flat background cannot present the ground state of the theory in the ultraviolet region.
11. It is shown, that the theory of gravity with a quadratic in the curvature and positive definite Euclidean action possesses a stable non-flat ground state and is asymptotically free both in the tensor sector and the conformal one. A physical interpretation of the nontrivial ground state as a condensate of conformal excitations, that is formed as a result of a phase transition, is proposed.
12. The effective potential, i.e., the off-shell one-loop effective action in arbitrary covariant gauge, and the Vilkovisky's effective action in the higher-derivative quantum gravity on De Sitter background, is calculated. The determinants of the operators of second and forth orders are obtained by means of the generalized ζ-function.
13. The gauge- and parametrization-independent unique effective equations for the background field, i.e., for the curvature of De Sitter space, are obtained. The perturbative solution of the effective equations, that gives the corrected value of the curvature of De Sitter background space due to quantum effects, is found.

References

1. L.F. Abbot, *The background field method beyond one loop*, Nucl. Phys. B, 1981, vol. 185, No 1, pp. 189–203.
2. S.L. Adler, *Einstein gravity as a symmetry breaking effect in quantum field theory*, Rev. Mod. Phys., 1982, vol. 54, No 3, pp. 729–766.
3. S.L. Adler, J. Lieberman and Y.J. Ng, *Regularization of the stress energy tensor and scalar particles propagating in general background metric*, Ann. Phys. (USA), 1977, vol. 106, pp. 279–321.
4. P. Amsterdamski, A. Berkin and D. O'Connor, b_8 *Hamidew coefficient for a scalar field*, Class. Quantum Grav., 1989, vol. 6, 1981–1991.
5. M.F. Atiyah, R. Bott and V.K. Patodi, *On the heat equation and the index theorem*, Invent. Math., 1973, vol. 19, pp. 279–330.
6. I.G. Avramidi, *Background field method in quantum theory*, Deposited at VINITI, No 1512-85 Dep., (Moscow: Moscow State Univ., 1984), 41 pp.
7. I.G. Avramidi, *Asymptotic behavior of gravity theory with higher derivatives*, Yadernaya Fiz., 1986, vol. 44, No 1(7), pp. 255–263.
8. I. G. Avramidi, *Covariant methods of studying the nonlocal structure of an effective action*, Yadernaya Fiz., 1989, vol. 49, pp. 1185–1192, [Russian]; Sov. J. Nucl. Phys., 1989, vol. 49, pp. 735–739 [English]
9. I.G. Avramidi, *Background field calculations in quantum field theory (vacuum polarization)*, Teoret. Matem. Fiz., 1989, vol. 79, pp. 219–231 [Russian]; Theor. Math. Phys., 1989, vol. 79, pp. 494–502 [English]
10. I.G. Avramidi, *The nonlocal structure of one-loop effective action via partial summation of asymptotic expansion*, Phys. Lett. B, 1990, vol. 236, pp. 443–449
11. I.G. Avramidi, *The covariant technique for calculation of the heat kernel asymptotic expansion*, Phys. Lett. B, 1990, vol. 238, pp. 92–97
12. I.G. Avramidi, *A covariant technique for the calculation of the one-loop effective action*, Nucl. Phys. B, 1991, vol. 355, pp. 712–754; Erratum: Nucl. Phys. B, 1998, vol. 509, pp. 557–558
13. I.G. Avramidi, *A new algebraic approach for calculating the heat kernel in gauge theories*, Phys. Lett. B, 1993, vol. 305, pp. 27–34
14. I.G. Avramidi, *The heat kernel on symmetric spaces via integrating over the group of isometries*, Phys. Lett. B, 1994, vol. 336, pp. 171–177

15. I.G. Avramidi, *Covariant algebraic calculation of the one-loop effective potential in non-Abelian gauge theory and a new approach to stability problem*, J. Math. Phys., 1995, vol. 36, pp. 1557–1571

16. I.G. Avramidi, *Covariant algebraic method for calculation of the low-energy heat kernel*, J. Math. Phys., 1995, vol. 36, pp. 5055–5070

17. I.G. Avramidi, *New algebraic methods for calculating the heat kernel and the effective action in quantum gravity and gauge theories*, in: *Heat Kernel Techniques and Quantum Gravity*, Ed. S.A. Fulling, Discourses in Mathematics and Its Applications, (College Station, Texas: Department of Mathematics, Texas A& M University, 1995), pp. 115–140

18. I.G. Avramidi, *A new algebraic approach for calculating the heat kernel in quantum gravity*, J. Math. Phys., 1996, vol. 37, pp. 374–394

19. I.G. Avramidi, *Covariant approximation schemes for calculation of the heat kernel in quantum field theory*, in: *Quantum Gravity*, Proc. VIth Moscow Int. Sem., (Singapore: World Scientific, 1997), pp. 61–78

20. I.G. Avramidi, *Nonperturbative methods for calculating the heat kernel*, in: Proc. Int. Conf. *Global Analysis, Differential Geometry and Lie Algebras*, Thessaloniki, Greece, Dec. 15-17, 1994, Ed. G. Tsagas, (Bucharest: Geometry Balcan Press, 1998), pp. 7–21

21. I.G. Avramidi, *Covariant techniques for computation of the heat kernel*, Rev. Math. Phys., 1999 vol. 11, No 8, pp. 947–980

22. I.G. Avramidi and A.O. Barvinsky, *Asymptotic freedom in higher-derivative quantum gravity*, Phys. Lett. B, 1985, vol. 159, No 4,5,6, pp. 269–274.

23. I.G. Avramidi and R. Schimming, *Heat kernel coefficients to the matrix Schrödinger operator*, J. Math. Phys., 1995, vol. 36, pp. 5042–5054

24. I.G. Avramidi and R. Schimming, *Algorithms for the calculation of the heat kernel coefficients*, in: *Quantum Field Theory under the Influence of External Conditions*, Ed. M. Bordag, Teubner-Texte zur Physik, Band 30, (Stuttgart: Teubner, 1996), pp. 150–162

25. N.H. Barth and S.M. Christensen, *Quantizing fourth-order gravity theories: The functional integral*, Phys. Rev. D, 1983, vol. 28, No 8, pp. 1876–1893.

26. A.O. Barvinsky and Y.V. Gusev, *Covariant technique for nonlocal terms of one loop radiative currents*, Russ. Phys. J., 1991, vol. 34, p. 858.

27. A.O. Barvinsky and Y.V. Gusev, *Covariant algorithms for one loop radiation currents in gauge theories and quantum gravity*, Class. Quant. Grav., 1992, vol. 9, p. 383.

28. A.O. Barvinsky, Y.V. Gusev, G.A. Vilkovisky and V.V. Zhytnikov, *The Basis of nonlocal curvature invariants in quantum gravity theory. (Third order.)*, J.Math.Phys., 1994, vol. 35, p. 3525

29. A.O. Barvinsky, Yu.V. Gusev, G.A. Vilkovisky and V.V. Zhytnikov, *Asymptotic Behaviour of the Heat Kernel in Covariant Perturbation Theory*, J. Math. Phys. 1994, vol. 35, pp. 3543–3559.

30. A.O. Barvinsky, Y.V. Gusev, G.A. Vilkovisky and V.V. Zhytnikov, *The One loop effective action and trace anomaly in four-dimensions*, Nucl. Phys. B, 1995, vol. 439, p. 561.

31. A.O. Barvinsky, Y.V. Gusev, V.V. Zhytnikov and G.A. Vilkovisky, *Covariant perturbation theory. 4. Third order in the curvature*, PRINT-93-0274 (University of Manitoba, 1993).

32. A.O. Barvinsky, Y.V. Gusev, V.V. Zhytnikov and G.A. Vilkovisky, *Asymptotic behaviors of one loop vertices in the gravitational effective action*, Class. Quant. Grav., 1995, vol. 12, p. 2157.

33. A.O. Barvinsky and G.A. Vilkovisky, *Divergences and anomalies for coupled gravitational and Majorana spin-1/2 fields*, Nucl. Phys. B, 1981, vol. 191, No 1, pp. 237–259.

34. A.O. Barvinsky and G.A. Vilkovisky, *The generalized Schwinger–De Witt technique and the unique effective action in quantum gravity*, Phys. Lett. B, 1983, vol. 131, No 4,5,6, pp. 313–318.

35. A.O. Barvinsky and G.A. Vilkovisky, *The generalized Schwinger–De Witt technique in gauge theories and quantum gravity*, Phys. Rep. C, 1985, vol. 119, No 1, pp. 1–74.

36. A.O. Barvinsky and G.A. Vilkovisky, *The generalized Schwinger–De Witt technique and unique effective action in quantum gravity*, in: *Quantum gravity*, Proc. IIIrd Sem. Quantum Gravity, Moscow 1984, Eds. M. A. Markov, V. A. Berezin and V. P. Frolov, (Singapore: World Sci. Publ., 1985), pp. 141–160.

37. A.O. Barvinsky and G.A. Vilkovisky, *Beyond the Schwinger–De Witt Technique: Converting Loops into Trees and In-In Currents*, Nucl. Phys. B, 1987, vol. 282, pp. 163–188.

38. A.O. Barvinsky and G.A. Vilkovisky, *The Effective Action In Quantum Field Theory: Two Loop Approximation*, in: Quantum Field Theory and Quantum Statistics, vol. 1, pp. 245–275.

39. A.O. Barvinsky and G.A. Vilkovisky, *Covariant Perturbation Theory. 2: Second Order In The Curvature. General Algorithms*, Nucl. Phys. B, 1990, vol. 333, p. 471.

40. A.O. Barvinsky and G.A. Vilkovisky, *Covariant Perturbation Theory. 3: Spectral Representations Of The Third Order Form-Factors*, Nucl. Phys. B, 1990, vol. 333, p. 512

41. F.A. Berezin, *Introduction to algebra and analysis with anticommuting variables*, (Moscow: Moscow State Univ. 1983), 208 pp.

42. N.D. Birrel and P.C.W. Davies, *Quantum fields in curved space*, (Cambridge: Cambridge Univ. Press, 1982)

43. N.D. Birrel and I.G. Taylor, *Analysis of interacting quantum field theory in curved space-time*, J. Math. Phys. (USA), 1980, vol. 21, No 7, pp. 1740–1760.

44. M.J. Booth, *HeatK: A Mathematica program for computing heat kernel coefficients*, Preprint JHU-TIPAC-98005, hep-th/9803113.

45. L.S. Brown, *Stress-tensor trace anomaly in a gravitational metric: Scalar fields*, Phys. Rev. D, 1977, vol. 15, No 6, pp. 1469–1483.

46. M. Brown, *Solutions of the wave equation in curved spacetime: Non-existence of the De Witt integral in De Sitter space-time*, Class. Quant. Grav., 1985, vol. 2, No 4, pp. 535–538.

47. L.S. Brown and J.P. Cassidy, *Stress-tensor trace anomaly in a gravitational metric: General theory, Maxwell fields*, Phys. Rev. D, 1977, vol. 15, pp. 2810–2829.

48. L.S. Brown and J.C. Collins, *Dimensional renormalization of scalar field theory in curved spacetime*, Ann. Phys. (USA), 1980, vol. 130, No 1, pp. 215–248.

49. T.S. Bunch, *BPHZ renormalization of field theory in curved spacetime*, Ann. Phys. (USA), 1981, vol. 131, No 1, pp. 118–148.

50. N.N. Bogolyubov and D.V. Shirkov, *Introduction to the theory of quantized fields*, (New York: Wiley, 1980), 620 pp.

51. D.G. Boulware, *Gauge dependence of the effective action*, Phys. Rev. D, 1981, vol. 23, No 2, pp. 389–396.

52. T.P. Branson, P.B. Gilkey and B. Ørsted, *Leading terms in the heat invariants*, Proc. Amer. Math. Soc., 1990, vol. 109, pp. 437–450.

53. I.L. Buchbinder, S.D. Odintsov and I.L. Shapiro, *Effective Action in Quantum Gravity*, (Bristol: IOP Publishing, 1992), 409 pp.

54. D.M. Capper, *A general gauge graviton loop calculation*, J. Phys. A: Gen. Phys., 1980, vol. 13, No 1, pp. 199–213.

55. D.M. Capper and M.J. Duff, *Trace anomalies in dimensional regularization*, Nuovo Cim. A, 1974, vol. 23, pp. 173–183.

56. D.M. Capper and M.J. Duff, *One-loop neutrino contribution to the graviton propagator*, Nucl. Phys. B, 1974, vol. 82, No 1, pp. 147–154.

57. D.M. Capper and M.J. Duff, *Conformal anomalies and renormalizability problem in quantum gravity*, Phys. Lett. A, 1975, vol. 53, No 5, pp. 361–362.

58. D.M. Capper, M.J. Duff and L. Halpern, *Photon corrections to the graviton propagator*, Phys. Rev. D, 1974, vol. 10, No 2, pp. 461–467.

59. D.M. Capper, J.J. Dulwich and R.M. Medrano, *The background field method for quantum gravity at two loops*, Nucl. Phys. B, 1985, vol. 254, No 3,4, pp. 737–746.

60. D.M. Capper, G. Leibrandt and M.R. Medrano, *Calculation of the graviton self-energy using dimensional regularization*, Phys. Rev. D, 1973, vol. 8, No 12, pp. 4320–4331.

61. D.M. Capper and A. Mac–Lean, *The background field method at two loops. A general gauge Yang–Mills calculation*, Nucl. Phys. B, 1982, vol. 203, No 3, pp. 413–422.

62. S.M. Christensen, *Vacuum expectation of the stress tensor in an arbitrary curved background: the covariant point separation method*, Phys. Rev. D, 1976, vol. 14, No 10, pp. 2490–2501.

63. S.M. Christensen, *Regularization, renormalization and covariant geodesic point separation*, Phys. Rev. D, 1978, vol. 17, No 4, pp. 946–963.

64. S.M. Christensen and M. J. Duff, *Axial and conformal anomalies for arbitrary spin in gravity and supergravity*, Phys. Lett. B, 1978, vol. 76, No 5, pp. 571–574.

65. S.M. Christensen and M.J. Duff, *New gravitational index theorems and super theorems*, Nucl. Phys. B, 1979, vol. 154, pp. 301–342.

66. S.M. Christensen and M.J. Duff, *Quantizing gravity with a cosmological constant*, Nucl. Phys. B, 1980, vol. 170, [FS1], No 3, pp. 480–506.

67. C.J.S. Clarke, *The Analysis of Space-Time Singularities*, Cambridge Lecture Notes in Physics, Vol. 1 (Cambridge: Cambridge University Press, 1993), 175 pp.

68. G. Cognola and S. Zerbini, *Heat Kernel Expansion In Geometric Fields*, Phys.Lett. B, 1987, vol. 195, p. 435.

69. G. Cognola and S. Zerbini, *Some Physical Applications Of The Heat Kernel Expansion*, Mod. Phys. Lett. A, 1988, vol. 3, p. 599.

70. V.I. Denisov and A.A. Logunov, *Does the gravitational radiation exist in general relativity?*, Teoret. Mat. Fiz., 1980, vol. 43, No 2, pp. 187–201.

71. V.I. Denisov and A.A. Logunov, *Does the general relativity have the classical Newtonian limit?*, Teoret. Mat. Fiz., 1980, vol. 45, No 3, pp. 291–301.

72. S. Deser, *Quantum gravitation: Trees, loops and renormalization*, in: *Quantum gravity*, Oxford Symp. 1974, Eds. C.J. Isham, R. Penrose and D.W. Sciama, (Oxford: Oxford Univ. Press, 1975), pp. 136–173.

73. S. Deser, M.J. Duff and C.J. Isham, *Nonlocal conformal anomalies*, Nucl. Phys. B, 1976, vol. 111, pp. 45–55.

74. S. Deser, H.S. Tsao and P. van Nieuwenhuizen, *One-loop divergences of the Einstein–Yang–Mills system*, Phys. Rev. D, 1974, vol. 10, No 10, pp. 3337–3342.

75. S. Deser and P. van Nieuwenhuizen, *Non-renormalizability of the quantized Einstein–Maxwell system*, Phys. Rev. Lett., 1974, vol. 32, pp. 245–247.

76. S. Deser and P. van Nieuwenhuizen, *One-loop divergences of quantized Einstein–Maxwell fields*, Phys. Rev. D, 1974, vol. 10, No 2, pp. 401–410.

77. S. Deser and P. van Nieuwenhuizen, *Non-renormalizability of the quantized Dirac–Einstein system*, Phys. Rev. D, 1974, vol. 10, No 2, pp. 411–420.

78. B.S. De Witt, *Quantum theory of gravity II. The manifestly covariant theory*, Phys. Rev., 1967, vol. 162, No 5, pp. 1195–1238.

79. B.S. De Witt, *Gravity: A universal regulator*, Phys. Rev. Lett., 1964, vol. 13, No 3, pp. 114–118.

80. B.S. De Witt, *Dynamical theory of groups and fields*, (New York: Gordon and Breach, 1965), 230 pp.

81. B.S. De Witt, *Quantum theory of gravity III. The application of the covariant theory*, Phys. Rev., 1967, vol. 162, No 5, pp. 1239–1256.

82. B.S. De Witt, *Quantum field theory in curved spacetime*, Phys. Rep. C, 1975, vol. 19, pp. 296–357.

83. B.S. De Witt, *Quantum gravity: New synthesis*, in: *General relativity*, Eds. S. Hawking and W. Israel, (Cambridge: Cambridge Univ. Press. 1979)

84. B.S. De Witt, *Gauge invariant effective action*, in: *Quantum gravity II*, Second Oxford Symp. 1980, Eds. C. J. Isham, R. Penrose and D. W. Sciama, (Oxford: Oxford Univ. Press, 1981), pp. 449–487.

85. B.S. De Witt, *The Spacetime Approach to Quantum Field Theory*, in: *Relativity, Groups and Topology II*, Eds. B.S. De Witt and R. Stora (Amsterdam: North Holland, 1984) p. 393

86. B.S. De Witt, *The effective action*, in: *Architecture of fundamental interactions at short distances*, Eds. P. Ramond and R. Stora (Amsterdam: North Holland, 1987), pp. 1023–1057

87. B.S. De Witt, *Supermanifolds*, (Cambridge: Cambridge University Press, 1992), 407 pp.

88. J.S. Dowker, *Single loop divergences in six dimensions*, J. Phys. A: Gen. Phys., 1978, vol. 10, pp. 63–69.

89. J.S. Dowker and R. Critchley, *Effective Lagrangian and energy-momentum tensor in de Sitter space*, Phys. Rev. D, 1976, vol. 13, No 12, pp. 3224–3232.

90. J.S. Dowker and R. Critchley, *Stress-tensor conformal anomaly for scalar, spinor and vector fields*, Phys. Rev. D, 1977, vol. 16, No 12, pp. 3390–3394.

91. M.J. Duff, *Covariant quantization*, in: *Quantum gravity*, Oxford Symp. 1974, Eds. C.J. Isham, R. Penrose and D.W. Sciama, (Oxford: Oxford Univ. Press, 1975), pp. 78–135.

92. M.J. Duff, *Observations of conformal anomalies*, Nucl. Phys. B, 1977, vol. 125, No 2, pp. 334–348.

93. M.J. Duff, *Inconsistency of quantum field theory in curved spacetime*, in: *Quantum gravity II*, Second Oxford Symp. 1980, Eds. C.J. Isham, R. Penrose and D.W. Sciama, (Oxford: Oxford Univ. Press, 1981), pp. 81–105.

94. M.J. Duff, *The cosmological constant in quantum gravity and supergravity*, in: *Quantum gravity II*, Second Oxford Symp. 1980, Eds. C.J. Isham, R. Penrose and D.W. Sciama, (Oxford: Oxford Univ. Press, 1981), pp. 488–500.

95. M. Duff, *Ultraviolet divergences in extended supergravity theories*, in: *Supergravity 1981*, Eds. S. Ferrara and J. Taylor, (Cambridge: Cambridge Univ. Press, 1982), p. 197.

96. A. Eddington, *The mathematical theory of relativity*, (Cambridge: Cambridge Univ. Press, 1924), 311 pp.

97. E. Elizalde, *Ten Physical Applications of Spectral Zeta-Functions*, Lecture Notes in Physics, New Series m: Monographs, Vol. m35 (Berlin: Springer-Verlag, 1995).

98. A. Erdelyi, W. Magnus, F. Oberhettinger and F.G. Tricomi, *Higher Transcendental Functions*, vol. I, (New York: McGraw–Hill, 1953).

99. G. Esposito, *Quantum Gravity, Quantum Cosmology and Lorentzian Geometries*, Lecture Notes in Physics, New Series m: Monographs, vol. m12, (Berlin: Springer-Verlag, 1994).

100. L.D. Faddeev, *The energy problem in the theory of gravitation*, Uspekhi Fiz. Nauk, 1982, vol. 136, pp. 435–457.

101. L.D. Faddeev and V.N. Popov, *Feynman diagrams for the Yang–Mils field*, Phys. Lett. B, 1967, vol. 25, No 1, pp. 29–30.

102. L.D. Faddeev and V.N. Popov, *Covariant quantization of the gravitational field*, Uspekhi Fiz. Nauk, 1973, vol. 111, No 3, pp. 427–450.

103. S. Ferrara and J. Taylor, Eds., *Supergravity 1981*, (Cambridge: Cambridge Univ. Press, 1982).

104. R.P. Feynman, *Quantum theory of gravitation*, Acta Phys. Pol., 1963, vol. 24, No 6, pp. 697–722.

105. V.A. Fock, *The proper time in classical and quantum mechanics*, Izvestiya of USSR Academy of Sciences, Physics, 1937, No 4,5, pp. 551–568.

106. E.S. Fradkin and I.V. Tjutin, *S-matrix for Yang–Mils and gravitational fields*, Phys. Rev. D, 1970, vol. 25, No 12, pp. 2841–2857.

107. E.S. Fradkin and A.A. Tseytlin, *Higher derivative quantum gravity: One-loop counterterms and asymptotic freedom*, Preprint No 70, P.N. Lebedev Physical Inst., Moscow, 1981, 59 pp.

108. E.S. Fradkin and A.A. Tseytlin, *Renormalizable asymptotically free quantum theory of gravity*, Phys. Lett. B, 1981, vol. 104, No 5, pp. 377–381.

109. E.S. Fradkin and A.A. Tseytlin, *Renormalizable asymptotically free quantum theory of gravity*, Nucl. Phys. B, 1982, vol. 201, No 3, pp. 469–491.

110. E.S. Fradkin and A.A. Tseytlin, *Asymptotic freedom in extended conformal supergravities*, Phys. Lett. B, 1982, vol. 110, No 2, pp. 117–122.

111. E.S. Fradkin and A.A. Tseytlin, *Asymptotically free renormalizable theory of gravity and supergravity*, in: *Quantum gravity*, Proc. IInd Sem. Quantum Gravity, Moscow 1981, (Moscow: Inst. Nuclear Research, 1983), pp. 18–27.

112. E.S. Fradkin and A.A. Tseytlin, *Conformal anomaly in Weyl theory and anomaly free superconformal theories*, Phys. Lett. B, 1984, vol. 134, No 3,4, pp. 187–193.

113. E.S. Fradkin and A.A. Tseytlin, *One-loop effective potential in gauge $O(4)$ supergravity and the problem of the Λ-term*, Nucl. Phys. B, 1984, vol. 234, No 2, pp. 472–508.

114. E.S. Fradkin and A.A. Tseytlin, *On the new definition of off-shell effective action*, Nucl. Phys. B, 1984, vol. 234, No 2, pp. 509–523.

115. E.S. Fradkin and G.A. Vilkovisky, *S-matrix for gravitational field II. Local measure. General relations. Elements of renormalization theory*, Phys. Rev. D, 1973, vol. 8, pp. 4241–4285.

116. E.S. Fradkin and G.A. Vilkovisky, *On renormalization of quantum field theory in curved space-time*, Preprint, Inst. Theoretical Physics, Univ. Bern, Bern, 1976, 144 pp.

117. E.S. Fradkin and G.A. Vilkovisky, *Quantization of relativistic systems with constraints. Equivalence of canonical and covariant formalism in quantum theory of gravitational field*, Preprint Th-2332, CERN, Geneva, 1977, 53 pp.

118. E.S. Fradkin and G.A. Vilkovisky, *Conformal off mass-shell extensions and elimination of conformal anomalies in quantum gravity*, Phys. Lett. B, 1978, vol. 73, No 2, pp. 209–213.

119. E.S. Fradkin and G.A. Vilkovisky, *Conformal invariance and asymptotic freedom in quantum gravity*, Phys. Lett. B, 1978, vol. 77, No 3, pp. 262–266.

120. V.P. Frolov, *Vacuum polarization near the black holes*, in: *Quantum theory of gravity*, Proc. IInd Sem. Quantum Gravity, Moscow 1981, (Moscow: Inst. Nuclear Research, 1983), pp. 176–187.

121. V.P. Frolov, *Physical effects in the gravitational field of the black holes*, Proc. P. N. Lebedev Physical Inst., 1986, vol. 169, pp. 3–131.

122. V.P. Frolov and G.A. Vilkovisky, *Spherically symmetric collapse in quantum gravity*, Phys. Lett. B, 1981, vol. 106, No 4, pp. 307–313.

123. S.A. Fulling, *Aspects of Quantum Field Theory in Curved Space-Time*, (Cambridge: Cambridge University Press, 1989).

124. S.A. Fulling, Ed, *Heat Kernel Techniques and Quantum Gravity* (*Discourses in Mathematics and Its Applications, No. 4*). (College Station: Texas A&M University, 1995).

125. S.A. Fulling and G. Kennedy, *The resolvent parametrix of the general elliptic linear diffrential operator: a closed form for the intrinsic symbol*, Trans. Amer. Math. Soc., 1988, vol. 310, pp. 583–617.

126. F. Englert, C. Truffin and R. Gastmans, *Conformal invariance in quantum gravity*, Nucl. Phys. B, 1976, vol. 117, pp. 407–432.

127. M.D. Freeman, *Renormalization of non-Abelian gauge theories in curved space-time*, Ann. Phys. (USA), 1984, vol. 153, No 2, pp. 339–366.

128. H. Georgi, *Unified theory of elementary particles*, Uspekhi Fiz. Nauk, 1982, vol. 136, pp. 286–316.

129. G.M. Gibbons, *Quantum field theory in curved space-time*, in: *General relativity*, Eds. S.W. Hawking and W. Israel, (Cambridge: Cambridge Univ. Press, 1979), pp. 639–679.

130. G.W. Gibbons and S.W. Hawking, Eds. *Euclidean Quantum Gravity* (Singapore: World Scientific, 1993), 586 pp.

131. G.W. Gibbons, S.W. Hawking and M.J. Perry, *Path integral and the indefiniteness of the gravitational action*, Nucl. Phys. B, 1978, vol. 138, No 1, pp. 141–150.

132. P.B. Gilkey, *The spectral geometry of Riemannian manifold*, J. Diff. Geom., 1975, vol. 10, pp. 601–618.

133. P.B. Gilkey, *Invariance Theory, the Heat Equation, and the Atiyah-Singer Index Theorem* (Boca-Raton: Chemical Rubber Company, 1995).

134. V.L. Ginzburg, D.A. Kirzhnits and A.A. Lyubushin, *On the role of quantum fluctuations of the gravitational field in general relativity*, Zhurnal Experiment. Teoret. Fiz., 1971, vol. 60, No 2, pp. 451–459.

135. M.H. Goroff and A. Sagnotti, *The ultraviolet behavior of Einstein gravity*, Nucl. Phys. B, 1986, vol. 266, pp. 709–736.

136. A.A. Grib, *Problems of the non-invariance of the vacuum in quantum field theory*, (Moscow: Atomizdat, 1978), 127 pp.

137. A.A. Grib, S.G. Mamaev and V.M. Mostepanenko, *Vacuum quantum effects in strong fields*, (St. Petersburg: Friedmann Laboratory Pub., 1994), 361 pp.

138. M.T. Grisaru, P. van Nieuwenhuizen and C.C. Wu, *Background-field method versus normal field theory in explicit examples: One-loop divergences in the S-matrix and Green's functions for Yang–Mils and gravitational field*, Phys. Rev. D, 1975, vol. 12, No 10, pp. 3203–3213.

139. L.P. Grishchuk and Ya.B. Zeldovich, *Complete cosmological theories*, in: *Quantum gravity*, Proc. IInd Sem. Quantum Gravity, Moscow 1981, (Moscow: Inst. Nuclear Research, 1982), pp. 39–48.

140. P. Günther and R. Schimming, *Curvature and spectrum of compact Riemannian manifolds*, J. Diff. Geom., 1977, vol. 12, 599–618.

141. V.P. Gusynin, *New Algorithm For Computing The Coefficients In The Heat Kernel Expansion*, Phys. Lett. B, 1989, vol. 225, p. 233.

142. V.P. Gusynin and V.V. Kornyak, *Symbolic computation of the heat kernel expansion on curved manifolds*, ITF-93-59E, Proc. 3rd Int. Workshop on Software Engineering, Artificial Intelligence and Expert Systems for High Energy and Nuclear Physics, Oberammergau, Germany, Oct. 4–8, 1993; to be published.

143. V.P. Gusynin and V.A. Kushnir, *Derivative Expansion For The One Loop Effective Action In Curved Space*, Sov. J. Nucl. Phys., 1990, vol. 51, p. 373.

144. V.P. Gusynin and V.A. Kushnir, *On diagonal heat kernel expansion in covariant derivatives in curved space*, Class. Quant. Grav., 1991, vol. 8, p. 279.

145. B. Hasslacher and E. Mottola, *Asymptotically free quantum gravity and black holes*, Phys. Lett. B, 1981, vol. 99, No 3, pp. 221–224.

146. S.J. Hathrell, *Trace anomalies and QED in curved space*, Ann. Phys. (USA), 1982, vol. 142, No 1, pp. 36–63.

147. S.J. Hathrell, *Trace anomalies and $\lambda\varphi^4$ theory in curved space*, Ann. Phys. (USA), 1982, vol. 149, No 1, pp. 136–197.

148. J. Hadamard, *Lectures on Cauchy's Problem in Linear Partial Differential Equations*, (New York, Yale University Press, 1923), 316 pp.

149. S.W. Hawking, *Zeta function regularization of path integrals in curved spacetime*, Comm. Math. Phys., 1977, vol. 55, pp. 133–148.

150. S. Hawking, *Path integrals applied to quantum gravity*, in: *General relativity*, Eds. S.W. Hawking and W. Israel, (Cambridge: Cambridge Univ. Press, 1979)

151. S. Hawking and J. Ellis, *The large-scale structure of space-time*, (Cambridge: Cambridge Univ. Press, 1973).

152. N.E. Hurt, *Geometric quantization in action: Applications of harmonic analysis in quantum statistical mechanics and quantum field theory*, (Dordrecht: Reidel, 1983).

153. S. Ichinose and M. Omote, *Renormalization using the background field method*, Nucl. Phys. B, 1982, vol. 203, No 2, pp. 221–267.

154. C.J. Isham, *Quantum gravity — an overview*, in: *Quantum gravity II*, Second Oxford Symp. 1980, Eds. C.J. Isham, R. Penrose and D.W. Sciama, (Oxford: Oxford Univ. Press, 1981), pp. 1–62.

155. C. Itzykson and J.-B. Zuber, *Quantum field theory*, (New York: McGraw–Hill, 1980).

156. D.D. Ivanenko and G.A. Sardanashvili, *Gravitation*, (Kiev: Naukova Dumka, 1985), 198 pp.

157. I. Jack and H. Osborn, *Two-loop background field calculations for arbitrary background fields*, Nucl. Phys. B, 1982, vol. 207, No 3, pp. 474–504.

158. I. Jack and H. Osborn, *Background field calculations in curved space-time (I). General formalism and application to scalar fields*, Nucl. Phys. B, 1984, vol. 234, No 2, pp. 331–364.

159. I. Jack, *Background field calculations in curved space-time (II). Application to a pure gauge theory*, Nucl. Phys. B, 1984, vol. 234, No 2, pp. 365–378.

160. J. Julve and M. Tonin, *Quantum gravity with higher derivative terms*, Nuovo Cim. B, 1978, vol. 46, No 1, pp. 137–152.

161. R.E. Kallosh, *Renormalization in non-Abelian gauge theories*, Nucl. Phys. B, 1974, vol. 78, No 2, pp. 293–312.

162. R.E. Kallosh, *Modified Feynman rules in supergravity*, Nucl. Phys. B, 1978, vol. 141, No 1,2, pp. 141–152.

163. R.E. Kallosh, O.V. Tarasov and I.V. Tjutin, *One-loop finiteness of quantum gravity off mass shell*, Nucl. Phys. B, 1978, vol. 137, No 1,2, pp. 145–163.

164. O.Y. Karmanov, *Effective action and spontaneous symmetry breaking in curved space-time*, in: *Problems of the theory of gravitation and elementary particles*, No 13, Ed. K.P. Stanyukovich, (Moscow: Energoatomizdat, 1982), pp. 46–57.

165. I.B. Khriplovich, *Gravitation and finite renormalizations in quantum electrodynamics*, Yadernaya Fiz., 1966, vol. 3, No 3, pp. 575–581.

166. T.W.B. Kibble, *Is semiclassical gravity theory viable?*, in: *Quantum gravity II*, Second Oxford Symp. 1980, Eds. C.J. Isham, R. Penrose and D.W. Sciama, (Oxford: Oxford Univ. Press, 1981), pp. 63–80.

167. N.P. Konopleva, Ed., *Quantum theory of gauge fields*, (Moscow: Mir, 1977), 432 pp.

168. D. Kramer, H. Stephani, M. MacCallum and E. Herlt, *Exact solutions of Einstein's field equations*, (New York: Cambridge University Press, 1980), 425 pp.

169. C. Lee, *Proper-time renormalization of multi-loop amplitudes in the background field method (I). Φ^4-theory*, Nucl. Phys. B, 1982, vol. 207, No 1, pp. 157–188.

170. S.C. Lee and P. van Nieuwenhuizen, *Counting of states in higher derivative field theories*, Phys. Rev. D, 1982, vol. 26, No 4, pp. 934–937.

171. A.D. Linde, *Inflation and Quantum Cosmology* (New York: Academic Press, 1990), 199 pp.

172. A.A. Logunov and V.N. Folomeshkin, *Does the radiation of the gravitational waves change the energy of a source in Einstein theory of gravitation?*, Teoret. Mat. Fiz., 1977, vol. 33, No 2, pp. 174–184.

173. S. Mandelstam, *Feynman rules for the gravitational fields from the coordinate-independent field-theoretic formalism*, Phys. Rev., 1968, vol. 175, No 5, pp. 1604–1623.

174. Y.I. Manin, Ed., *Geometric ideas in physics*, (Moscow: Mir, 1983), 240 pp.

175. H.P. McKean and I. Singer, *Curvature and the eigenvalues of the Laplacian*, J. Diff. Geom., 1967, vol. 1, pp. 43–69.

176. S. Minakshisundaram, *Eigenfunctions on Riemannian manifolds*, J. Indian Math. Soc., 1953, vol. 17, pp. 158–165

177. S. Minakshisundaram and A. Pleijel, *Some properties of the eigenfunctions of the Laplace operator on Riemannian manifolds.* Can. J. Math., 1949, vol. 1, pp. 242–256.

178. C.W. Misner, K.S. Thorne and J.A. Wheeler, *Gravitation*, (San Francisco: Freeman, 1973).

179. F.H. Molzahn, T.A. Osborn and S.A. Fulling, *Gauge Invariant Asymptotic Expansion Of Schrodinger Propagators On Manifolds*, Annals Phys., 1990, vol. 204, p. 64.

180. D.F. Neville, *Gravity Lagrangian with ghost-free curvature-squared terms*, Phys. Rev. D, 1978, vol. 18, No 10, pp. 3535–3541.

181. D. Neville, *Conformal divergences and spacetime foam in* $R+R^2$ *theory*, Phys. Rev. D, 1982, vol. 26, No 10, pp. 2638–2644.

182. N.K. Nielsen, *Ghost counting in supergravity*, Nucl. Phys. B, 1978, vol.140, No 3, pp. 499–509.

183. N.K. Nielsen, H. Römer and B. Schröer, *Anomalous currents in curved space*, Nucl. Phys. B, 1978, vol. 136, No 3, pp. 475–492.

184. M. Nouri–Moghadam and J.G. Taylor, *One-loop divergences for Einstein-charged meson system*, Proc. Roy. Soc. London, 1975, vol. 344, No 1636, pp. 87–99.

185. M. Nouri–Moghadam and J.G. Taylor, *Ghost elimination in quantum gravity*, J. Phys. A: Gen. Phys. , 1976, vol. 9, No 1, pp. 59–71.

186. M. Nouri–Moghadam and J.G. Taylor, *Ghost killing in quantized Einstein-Maxwell theory*, J. Phys. A: Gen. Phys., 1976, vol. 9, No 1, pp. 73–76.

187. L. Parker, *Aspects of quantum field theory in curved space-time*, in *Recent Developments in Gravitation*, Proc. NATO Advanced Institute, Cargese 1978, Eds. Maurice Levy and S. Deser (Plenum Publishing, 1979).

188. I. Polterovich, *A commutator method for computation of heat invariants*, Preprint math.DG/9805047.

189. I. Polterovich, *Heat invariants of Riemannian manifolds*, Preprint match.DG/9905073.

190. V.N. Ponomarev, A.O. Barvinsky and Yu.N. Obukhov, *Geometro-dynamical methods and the gauge approach in the theory of gravitational interactions*, (Moscow: Energoatomizdat, 1984), 167 pp.

191. V.N. Popov, *Functional integrals in quantum field theory and statistical physics*, (Boston: Reidel, 1983), 299 pp.

192. R. Rajaraman, *Solitons and instantons: an introduction to solitons and instantons in quantum field theory*, (Amsterdam: North-Holland, 1982), 409 pp.

193. P. Ramond, *Field theory: A modern primer*, (Massachusetts: Benjamin, Reading, 1981).

194. R.J. Riegert, *A nonlocal action for the trace anomaly*, Phys. Lett. B, 1984, vol. 134, No 1, pp. 56–60.

195. T.V. Ruzmaikina and A.A. Ruzmaikin, *Quadratic corrections to Lagrangian density of the gravitational field and singularity*, Zhurnal Experiment. Teoret. Fiz., 1969, vol. 57, No 8, pp. 680–685.

196. T. Sakai, *On eigenvalues of Laplacian and curvature of Riemannian manifold*, Tohoku Math. J., 1971, vol. 23, pp. 589–603.

197. A. Salam, *Calculation of the renormalization constants*, in: *Quantum gravity and topology*, Ed. D. D. Ivanenko, (Moscow: Mir, 1973), pp. 180–215.

198. A. Salam and J. Strathdee, *Remarks on high-energy stability and renormalizability of gravity theory*, Phys. Rev. D, 1978, vol. 18, No 12, pp. 4480–4485.

199. R. Schimming, *Lineare Differentialoperatoren und die Methode der Koinzidenzwerte in der Riemannschen Geometrie*, Beitr. z. Analysis, 1981, vol. 15, pp. 77–91.

200. R. Schimming, *Spektrale Geometrie und Huygenssches Prinzip für Tensorfelder und Differentialformen I*, Z. f. Analysis u. ihre Anw. 1982, vol. 2, pp. 71–95.

201. R. Schimming, *Laplace-like linear differential operators with a logarithm-free elementary solution*, Math. Nachr., 1990, vol. 147, 217–246.

202. R. Schimming, *Calculation of the heat kernel coefficients*, in: *Analysis, Geometry and Groups. A Riemann Legacy Volume*, Eds. H. M. Srivastava and Th. M. Rassias, (Palm Harbor: Hadronic Press, 1993), p. 627–656

203. J.S. Schwinger, *On gauge invariance and vacuum polarization*, Phys. Rev., 1951, vol. 82, No 5, pp. 664–679.

204. J.S. Schwinger, *The theory of quantized fields V.*, Phys. Rev., 1954, vol. 93, No 3, pp. 615–628.

205. R.T. Seeley, *Complex powers of an elliptic operator*, Proc. Symp. Pure Math., 1967, vol. 10, pp. 288–307.

206. E. Sezgin and P. van Nieuwenhuizen, *Renormalizability properties of antisymmetric tensor fields coupled to gravity*, Phys. Rev. D, 1980, vol. 22, No 2, pp. 301–307.

207. A.A. Slavnov and L.D. Faddeev, *Gauge fields, introduction to quantum theory*, (Mass.: Addison-Wesley, 1991), 217 pp.

208. K.S. Stelle, *Renormalization of higher-derivative quantum gravity*, Phys. Rev. D, 1977, vol. 16, No 4, pp. 953–969.

209. K.S. Stelle, *Classical gravity with higher derivatives*, Gen. Rel. Grav., 1978, vol. 9, No 4, pp. 353–371.

210. A.G. Sveshnikov and A.N. Tikhonov, *The theory of functions of a complex variable*, (Moscow: Mir, 1971), 311 pp.

211. J.L. Synge, *Relativity: The general theory*, (Amsterdam: North-Holland, 1960).

212. G. t'Hooft and M. Veltman, *One-loop divergences in the theory of gravitation*, Ann. Inst. H. Poincare, 1974, vol. 20, No 1, pp. 69–94.

213. E. Tomboulis, *$1/N$-expansion and renormalization in quantum gravity*, Phys. Lett. B, 1977, vol. 70, No 3, pp. 361–364.

214. E. Tomboulis, *Renormalizability and asymptotic freedom in quantum gravity*, Phys. Lett. B, 1980, vol. 97, No 1, pp. 77–80.

215. E. Tomboulis, *Unitarity in higher-derivative quantum gravity*, Phys. Rev. Lett., 1984, vol. 52, No 14, pp. 1173–1176.

216. D.J. Toms, *Renormalization of interacting scalar field theory in curved space-time*, Phys. Rev. D, 1982, vol. 26, No 10, pp. 2713–2729.

217. R. Utiyama and B. S. De Witt, *Renormalization of a classical gravitational field interacting with quantized matter fields*, J. Math. Phys. (USA), 1962, vol. 3, No 4, pp. 608–618.

218. A.I. Vainshtein, V.I. Zakharov, V.A. Novikov and M.A. Shifman, *Calculations in external fields in QCD: The operator approach*, Yadernaya Fiz. 1984, vol. 39, No 1, pp. 124–137.

219. A.E. van de Ven, *Explicit Counter Action Algorithms In Higher Dimensions*, Nucl. Phys. B, 1985, vol. 250, p. 593.

220. A.E. van de Ven, *Index-free heat kernel coefficients*, Class. Quant. Grav., 1998, vol. 15, p. 2311.

221. P. van Nieuwenhuizen, *Supergravity*, Phys. Rep., 1981, vol. 68, No 4, pp. 189–398.

222. A. van Proeyen, *Gravitational divergences of the electromagnetic interactions of massive vector particles*, Nucl. Phys. B, 1980, vol. 174, No 1, pp. 189–206.

223. G.A. Vilkovisky, *The Gospel according to De Witt*, in: *Quantum Gravity*, Ed. S. Christensen, (Bristol: Hilger, 1983), pp. 169–209.

224. G.A. Vilkovisky, *The unique effective action in quantum field theory*, Nucl. Phys. B, 1984, vol. 234, pp. 125–137.

225. G.A. Vilkovisky and V.P. Frolov, *Spherically symmetric collapse in quantum gravity*, in: *Quantum theory of gravity*, Proc. IInd Sem. Quantum Gravity, Moscow 1981, (Moscow: Inst. Nuclear Research, 1983), pp. 155–169.

226. A.A. Vladimirov and D.V. Shirkov, *Renormalization group and ultraviolet asymptotics*, Uspekhi Fiz. Nauk, 1979, vol. 129, No 3, pp. 407–441.

227. B.L. Voronov and I.V. Tjutin, *On renormalization of the Einstein gravity*, Yadernaya Fiz., 1981, vol. 33, No 6, pp. 1710–1722.

228. B.L. Voronov and I.V. Tjutin, *On the renormalization of R^2-gravity*, Yadernaya Fiz., 1984, vol. 39, No 4, pp. 998–1010.

229. S. Weinberg, *Ultraviolet divergences in quantum gravity theories*, in: *General relativity*, Eds. S.W. Hawking and W. Israel, (Cambridge: Cambridge Univ. Press, 1979)

230. H. Weyl, *Raum, Zeit, Materie*, (Berlin: Springer, 1923), 333 pp.

231. H. Widom, *A complete symbolic calculus for pseudodifferential operators*, Bull. Sci. Math., 1980, vol. 104, pp. 19–63.

232. S. Yajima, *Evaluation of heat kernel in Riemann–Cartan space*, Class. Quantum Grav., 1996, vol. 13, 2423–2436.

233. S. Yajima, *Evaluation of the heat kernel in Riemann–Cartan space using the covariant Taylor expansion method*, Class. Quantum Grav., 1997, vol. 14, 2853–2868.

Notation

We use the unit system $\hbar = c = G = 1$ (except for some expressions where these quantities are contained explicitly) and the notation and the sign conventions of the book [178]. The local coordinates of the space–time are labeled by the Greek indices, $\mu = 0, 1, \dots, n-1$. A summation is always assumed over repeated upper and lower Greek indices. The signature of the space–time metric is chosen to be $(- + \cdots +)$. The Riemann curvature tensor, the Ricci tensor and the scalar curvature are defined by $[\nabla_\mu, \nabla_\nu]u^\alpha = R^\alpha{}_{\beta\mu\nu}u^\beta$, $R_{\mu\nu} = R^\alpha{}_{\mu\alpha\nu}$ and $R = g^{\mu\nu}R_{\mu\nu}$. The commutator of covariant derivatives of a field φ is denoted by $[\nabla_\mu, \nabla_\nu]\varphi = \mathcal{R}_{\mu\nu}\varphi$. The D'Alambert (or Laplace) operator is defined by $\Box = g^{\mu\nu}\nabla_\mu\nabla_\nu$.

We use the standard convention that a field index 'A' in the exponent of (-1) is equal to 0 for bosonic indices and to 1 for the fermionic ones [41, 87], i.e., $(-1)^A = 1$ for a bosonic A and $(-1)^A = -1$ for a fermionic A. Also, we use De Witt's condensed notation [80, 83]. That is a small Latin index 'i' denotes the mixed set of indices (A, x), where x labels the space-time point, i.e., $\varphi^i = \varphi^A(x)$. The combined summation-integration is assumed over repeated upper and lower small Latin indices, i.e., $\varphi_{1\,i}\varphi_2^i \equiv \int d^n x \varphi_{1\,A}(x)\varphi_2^A(x)$.

The symmetrization over tensor (or field) indices is denoted by parenthesis and anti-symmetrization by square brackets; the indices that are excluded from symmetrization or anti-symmetrization are separated by vertical lines.

Further, the square brackets denote the coincidence limit of a two-point function $[f(x, x')] \equiv \lim_{x \to x'} f(x, x')$. The commutator of operators or matrices is also denoted by square brackets, $[A, B] = AB - BA$, but this usually does not cause any misunderstanding. The anti-commutator is denoted by $[A, B]_+ = AB + BA$. The identity matrix is denoted by $\hat{1}$. The square brackets $[x]$ also denote the integer part of a real number x.

Index

Abel–Plan summation, 117
Action, 9, 78
– classical, 13, 18, 77, 85
– Euclidean, 5, 7, 102, 106, 109–112
– gravitational, 4
– of higher-derivative quantum gravity, 85
Affine parameter, 21, 22
Analytical continuation, 17, 19, 52, 53, 59, 60, 65, 66, 117
Anomaly, 3
– conformal, 68
Anti-commutator, 39
Asymptotic expansion, 6, 20, 34, 48, 51, 52, 70, 74
– of effective action, 48, 76
– of evolution function, 51
– of Green function, 14
– of transfer function, 34, 68, 69, 73
– of transfer function in De Sitter space, 73
Asymptotic freedom, 5, 7, 102–105, 107, 108
Asymptotic states, 107

Back-reaction, 2
Background field, 2, 3, 10–12, 51, 53, 55, 60, 74, 79, 80, 108
– on-shell, 108
– reparametrization invariant, 81
Background field gauge, 80
Background field method, 1, 5
Background metric, 48, 88
– flat, 99
Bernoulli numbers, 71
Bernoulli polynomials, 70
Bessel function, 74
Bianci identity, 61
Black hole, 48
Borel function, 52, 53, 73, 74
Borel series, 53
Borel sum, 6, 52, 53
Borel summation, 52
Boundary conditions, 10, 14
– causal, 65
– for formfactors, 59, 65
Branching point, 59, 65, 74

Causal structure of space-time, 1
Causality, 3, 4, 10
Change of field variables, 109
Characteristic length scale, 17
Christoffel symbols, 22, 81, 83, 84, 100, 114
Chronological ordering, 10
Co-differentiation, 46
Coincidence limit, 16, 21, 22, 24
– of derivatives of parallel displacement operator, 24
– of derivatives of world function, 24, 25
– of De Witt coefficients, 34
– of Green function, 6, 17, 19, 20, 59, 97
– of symmetrized covariant derivatives, 36
– of transfer function, 15, 68, 72
Commutator of covariant derivatives, 22, 26, 46, 91, 96
Completeness of eigenfunctions, 24
Complex power of differential operator, 116
Compton wave length, 3, 17
Condensed De Witt's notation, 9
Configuration space, 77, 78, 80, 84, 111, 114
– complex, 106
– physical, 78, 84
Conformal condensate, 107
Conformal invariance, 68
Conformal mode, 107, 108
Conformal sector
– of Einstein gravity, 106

– of higher-derivative quantum gravity, 5, 7, 107
Conformally invariant models, 108
Conjugate points, 15
Connection, 5, 14, 22, 82, 83
– affine, 22
– gauge-invariant, 80
– in configuration space, 12, 81, 82, 115
– non-metric, 83, 98
Constraints, 1, 109
– first type, 77
Coordinate transformation, 14, 22
Cosmological constant, 4, 85, 106–108
Cosmological scenario, 123
Counter-terms, 3, 18
Coupling constant, 5, 7, 18, 101
– conformal, 85, 102, 103, 105, 107
– cosmological, 85, 103–105
– effective (running), 101
– Einstein, 85, 101, 102, 105, 106
– essential, 101
– matter, 104, 105
– non-essential, 18, 101, 102
– of scalar field with gravity, 46, 104
– renormalized, 18
– topological, 85, 102, 105
– Weyl, 85, 102, 105
Covariance, 53
– of effective action, 80
Covariant Fourier components, 55
Covariant derivatives, 14, 22
– symmetrized, 24, 42
Covariant Fourier components, 55
Covariant Fourier integral, 5, 25, 26, 54
– for Dirac delta-function, 27
– for formfactors, 57
– for Green function, 54
– for propagator, 107
Covariant Taylor series, 5, 23, 24, 26, 28, 30, 31, 36, 83
– for De Witt coefficients, 34
– in configuration space, 81
Curvature, 31, 36, 42
– covariantly constant, 68
– of De Sitter space, 109, 122
– of space-time, 96
Curved space, 5

D'Alambert operator, 14
Degrees of freedom, 116
– of higher-derivative quantum gravity, 116
Determinant
– of differential operator, 7, 79, 116, 118
– of ghost operator, 113, 115
– of integro-differential operator, 98
De Sitter space, 7, 33, 69, 94, 108–110, 112, 114, 123
De Witt coefficients, 3–6, 16, 34–36, 53, 55, 56, 68, 69
– A_3, 42
– $A_{n/2}$, 66
– $[a_3]$, 6, 37, 38, 42
– $[a_4]$, 6, 37, 38, 42, 46, 48
– $[a_{(n-2)/2}]$, 60
– for spinor field, 48
– in De Sitter space, 69, 71
– in first order in background fields, 57
– in second order in background fields, 61
De Witt projector, 82
Diagrammatic technique, 5, 11, 13, 36, 37
– covariant, 55
– for De Witt coefficients, 55, 56
– for effective action, 81
– of perturbation theory, 1
Diffeomorphism group, 1, 9, 84
Differential operator, 3, 14, 24, 25, 35, 84, 88, 116–119, 121, 122
– constrained, 111
– elliptic, 3
– minimal, 14, 100
– non-minimal, 91
Dimension
– anomalous, 18, 102
– background, 53, 55, 95, 96
– of coupling constant, 18
– of space-time, 92
– physical, 19
Dirac delta-function, 11, 13, 27, 78, 84
Dirac matrices, 46
Disc of convergence, 52, 53
Distribution, 17

Effective action, 1–5, 11–13, 48, 49, 64–68, 75, 76, 78–81, 93, 95, 98, 100, 101, 104, 109, 116, 119–122
– gauge-invariant, 2, 79, 80
– in De Witt gauge, 114, 115
– in minimal gauge, 105, 106
– in second order in background fields, 63, 65
– in two dimensions, 67, 68
– non-analytic, 6
– non-local, 3, 6, 48, 65

- of massless field, 66
- off-shell, 2, 12, 79, 80, 82
- on De Sitter space, 7, 108, 112, 113, 121
- on-shell, 12, 82, 94, 116, 121
- one-loop, 3, 6, 13, 15, 19, 20, 46, 47, 61, 74, 79, 85, 93–95, 108, 119
- renormalized, 4, 18, 42, 48, 64, 67, 74, 101
- reparametrization invariant, 2, 12, 81
- Vilkovisky's, 2, 6, 7, 12, 81–83, 98–101, 105, 106, 114–116, 120, 121
Effective equations, 2, 12, 80, 109, 122
- Vilkovisky's, 7
Effective potential, 7, 108
- of higher-derivative quantum gravity, 108
- one-loop, 108
- Vilkovisky's, 7
Eigenfunctions, 23, 24, 28, 34
- dual, 23, 26
- orthonormal, 24
Eigenmatrix, 31
Eigenvalue, 23, 116
Einstein equations, 1, 48
- in two dimensions, 67
Einstein gravity, 83, 115
Energy-momentum tensor, 15, 48
- effective, 48
Entire function, 53, 57, 63, 68, 73, 74
Equations of motion, 85
- classical, 12, 77, 108
Euclidean action of higher-derivative quantum gravity, 106, 107
Euclidean sector of space-time, 55, 109
Euler beta-function, 57
Euler characteristic, 107, 109
Euler constant, 20
Euler gamma-function, 58, 70, 71
Evolution equation, 14
Evolution function, 14, 15, 51
Expansion
- covariant, 5
- in background dimension, 3, 95
- in eigenfunctions, 34
- in extremal, 100, 119
- near mass shell, 95
Exterior derivative, 46
Extremal, 77, 85, 95–99, 109

Field components
- bosonic, 9
- contravariant, 9
- covariant, 9
- fermionic, 9
Field configurations, 77
Field renormalization constant, 18, 102
Field variables
- anti-commuting, 79
- constrained, 110, 111, 115
- dynamical, 77
- gauge-invariant, 78, 110
- Grassmanian, 9
- group, 78
- non-physical, 83, 110, 111
- unconstrained, 110
Finite part of effective action, 121
Flat space expansion, 107
Formfactors, 6, 58, 59, 63, 65
Functional derivative, 68, 83
- covariant, 81, 98
- left, 10
- right, 11
Functional determinant, 47, 119
- ζ-function regularized, 117
- of minimal differential operator, 47
Functional formulation of quantum field theory, 5
Functional integral, 10, 12, 78, 81, 106, 107
- Gaussian, 12, 79
Functional Legendre transform, 11
Functional measure, 10, 12, 13, 78, 115
Functional superdeterminant, 13
Functional supertrace, 13
Functional trace, 116
- of transfer function, 62

Gauge condition, 77–80, 101, 105, 109
- covariant, 5
- De Donder–Fock–Landau, 88, 115
- De Witt, 79, 88, 95, 99, 112, 113
- harmonic, 88, 98, 115
- minimal, 88, 92, 97, 100
- orthogonal, 83, 114
Gauge field theory, 9, 12
- non-Abelian, 83
Gauge fixing, 2, 9, 79, 82
Gauge fixing operator, 79
Gauge fixing parameters, 93
Gauge fixing tensor, 93, 96
Gauge fixing term, 112
Gauge group, 1, 77, 78, 80, 84
Gauge invariance, 77, 82
Gauge parameters, 95, 113, 114
- minimal, 100
Gauge transformation, 77, 80, 82

- of constrained fields, 110
- of unconstrained fields, 110
Gell-Mann–Low beta-functions, 3, 18, 101, 104, 105
General relativity, 1
Generating functional, 1, 10, 11
Generators of gauge transformations, 77, 78, 80, 82, 84
Geodesic, 15, 21–23
- in configuration space, 80
Geodesic distance, 15
Ghost fields, 4, 9
- Faddeev–Popov, 79
- Nielsen–Kallosh, 79
Ghost operator, 88, 113, 115
- on De Sitter space, 112
Ghost propagators, 96
Ghost states, 107
Gravitational field, 1, 4, 6, 46, 48, 49, 84, 105
- background, 2, 3
- linearized, 2
Gravitational loop, 2
Gravitational propagator, 96
Graviton, 116
Green function, 2, 3, 12–15, 17, 19, 47, 49, 51, 53, 55, 59, 60, 70, 71, 73, 96, 97, 99
- causal (Feynman), 15, 55, 58, 59
- conformally invariant, 6, 60
- exact, 54
- free, 54
- in De Sitter space, 69
- in first order in background fields, 58
- of massless field, 59
- of matter field, 46
- of minimal operator, 46
- renormalized, 58, 60
Ground state, 117
- of higher-derivative quantum gravity, 107
Group parameter, 77
Group variables, 78

Heat kernel, 14
High-energy asymptotics, 6, 68, 101
- of conformal coupling, 7
- of conformal sector, 109
- of coupling constants, 6, 101
- of formfactors, 6, 59, 60, 65, 67
- of Green function, 51
- of higher-derivative quantum gravity, 4, 5, 18, 104
Higher-derivative quantum gravity, 2, 4–7, 12, 83, 85, 101, 105–108, 116
- conformally invariant, 5
Hypersurface, 33

Imaginary part
- of effective action, 48, 65, 116, 122
- of formfactors, 6, 59, 65
In-region, 10
In-vacuum, 10
Infinitesimal diffeomorphism, 84
Infinitesimal imaginary part of mass, 15
Infinitesimal renormalization, 18
Infrared divergences, 68
- of effective action, 67
- of Green function, 60
Initial condition, 22, 31, 32
Instability
- of flat space, 107
- of ground state, 117

Jacobi identity, 77
Jacobian, 119, 120, 122
- functional, 81, 112, 113, 115

Killing equations, 81, 84

Laplace operator, 111, 116, 117
- constrained, 111
Levi–Civita tensor, 85
Lie algebra, 77
Light-cone, 3
Light-cone singularities of Green function, 15, 17
Loop diagram, 12
Loop expansion, 2, 12
Low-energy asymptotics, 68
- of higher-derivative quantum gravity, 107

Many-point Green functions, 11, 18
- off-shell, 18, 79
Mass of quantum field, 46
Mass shell, 4, 12, 18, 77, 79, 94, 98, 101, 108, 109, 111, 114, 120, 122
Matrix algorithm for De Witt coefficients, 36
Matrix elements of differential operator, 24, 25, 35, 36, 38, 56
- dimensionless, 56
Matter fields, 1, 4, 5, 7, 104
- background, 2
Matter loop, 2
Mean field, 10

Method of covariant expansions, 5, 34
Metric
- gauge-invariant, 81
- of configuration space, 9, 12, 13, 81, 82, 84, 99, 115
- of space-time, 14, 37, 84, 108
- pseudo-Riemannian, 84
Metric theory of gravity, 83
Multiplicativity of functional determinant, 13
Multiplicity of eigenvalues, 116

Nöther identity, 77, 85, 94
Negative modes, 116, 121, 122
Non-renormalizability, 108
- of Einstein quantum gravity, 4
Normalization
- of effective action, 64, 67
- of formfactors, 58, 60, 64, 67
Number of modes, 117, 118

Off-shell renormalizability of higher-derivative quantum gravity, 101
One-loop approximation, 2–4, 81, 83, 109, 112, 114, 123
One-loop counter-terms in higher-derivative quantum gravity, 4, 5
One-loop divergences, 37, 85
- of higher-derivative quantum gravity, 6
Orbit, 77, 78, 82
Orbit space, 77, 78, 80, 82, 84
Orthonormality of eigenfunctions, 24
Out-region, 10
Out-vacuum, 10

Parallel displacement operator, 15, 21, 22, 24, 26
Parallel transport equation, 22
Parameters of gauge transformation, 84
Parametrization of quantum field, 2, 12, 79, 80, 84, 98, 101, 105, 109
Partial summation, 51, 55, 65, 76
- of asymptotic series, 17
- of Schwinger–De Witt expansion, 6, 65
Particle creation, 2, 17
Path integral, 10
Perturbation theory, 1, 2, 4, 12, 17, 51, 77, 81, 83, 85, 99, 107, 122
- covariant, 1
- linearized, 4
Phase transition in conformal sector, 107
Planck constant, 12, 17
Planck energy, 2
Planck mass, 122
Point transformations, 13
Potential term, 14, 68, 69
Propagator, 2, 11, 96, 97
- bare, 13
- exact, 1
- Feynman, 15
- free, 100
- gravitational, 107
- one-point, 10–12
Proper time, 3, 17
Proper time method, 3, 5, 14, 51
Pseudo-Euclidean region, 6, 59, 65

Quadratic part of action, 115
Quantization
- canonical, 10, 78
- of gauge theories, 6
Quantum corrections, 1, 12, 80
- on De Sitter space, 121
- to background curvature, 7
Quantum effects, 1–3
- gravitational, 123
Quantum field, 2, 46, 48, 79, 80, 101, 109
- conformal, 110, 114
- conformally invariant, 68
- gauge, 77
- low-spin, 7, 104, 105
- massive, 4, 6, 37, 46
- massive scalar, 116
- massive tensor, 116
- massless, 3, 6, 17, 19, 49, 59, 60, 65, 67, 104
- massless tensor, 116
- matter, 2, 105
- physical, 99, 116
- scalar, 46
- spinor, 46, 105
- tensor, 114
- transverse traceless tensor, 110
- vector, 46
- vector ghost, 114
Quantum field theory, 12
- non-Abelian gauge, 1
- renormalizable, 17
- unified, 1
Quantum gravity (*see also Higher-derivative quantum gravity*), 1, 4, 5, 108

– Einstein, 2, 3
– one-loop, 2, 4
– one-loop Einstein, 4
– two-loop Einstein, 4

Radiative corrections, 4
Radius of convergence, 52, 53
Recurrence relations for De Witt coefficients, 16, 34
Regularization, 5, 15, 17, 18, 51
– ζ-function, 3, 19
– analytical, 3, 19
– covariant, 3
– cut-off, 19
– dimensional, 3, 19, 20, 66, 91
– Pauli–Villars, 19
Regularizing function, 19
Regularizing parameter, 18, 19
Renormalization, 5, 15, 17, 18, 51
– finite, 20
– non-minimal, 19
Renormalization group, 5, 18, 101
Renormalization group equations, 6, 101, 104, 105
Renormalization parameter, 18, 19, 60, 67, 101, 119
Renormalization point, 18
Reparametrization
– of orbit space, 80
– of quantum field, 12
Reparametrization invariance, 80, 82
Representation of gauge group
– adjoint, 80
Representation of Lie algebra, 78

Scalar product, 9, 23
Scaling of functional determinant, 117
Scatering matrix (or S-matrix), 1, 4, 12, 18, 79, 80, 82, 85, 108
Schwinger average, 10
Schwinger–De Witt expansion, 53, 59, 65, 76
– in De Sitter space, 70
Schwinger–De Witt representation, 19
– of Green function, 47, 51, 73
Schwinger–De Witt technique, 3, 6, 91
Second variation of action, 2, 13, 85, 99, 111
Semi-classical approximation, 2, 6, 17, 51
Small distance behavior of Green function, 17
Source, 11
– classical, 10
Space-time, 1, 10, 14, 37, 55, 84, 96, 108, 109
– pseudo-Euclidean, 84
– topologically non-trivial, 107
Space-time foam, 107
Space-time singularities, 1
Spectrum of Laplace operator, 111
Spectrum of perturbation theory, 107
Spin of quantum field, 46
Spontaneous creation of De Sitter space, 123
Stability
– of De Sitter space, 111
– of flat space, 102, 108
Structure constants of gauge group, 77
Summation
– of asymptotic series, 17, 73
– of leading derivatives, 3
– of radiative corrections, 4
Summation-integration, 10
Superdeterminant, 13
Supergravity, 4
– conformal, 5
Supersymmetry conditions, 9
Symbolic computation, 6
– of De Witt coefficients, 4
Symmetric space, 68
Symmetry breaking, 2

Tachyon, 5, 107
Tangent space, 14, 53
Tangent vector, 21
Tensor sector of higher-derivative quantum gravity, 5, 7
Threshold, 6
Topological term, 85
Transfer equation, 16
Transfer function, 15, 17, 34, 51, 58, 62, 68–70, 72–74
– in De Sitter space, 73, 74
– in first order in background fields, 57
Tree diagrams, 11, 12
Two-point functions, 15, 80

Ultraviolet divergences, 3–6, 13, 17–19, 51, 93, 96
– R^2, 108
– conformal, 5
– of "Vilkovisky's" effective action, 6
– of coincidence limit of Green function, 100
– of determinant of differential operator, 91, 100

– of effective action, 85, 93–96, 98, 100, 101, 104, 120
– of Green function, 60, 97
– of higher-derivative quantum gravity, 5
– off-shell, 6
– two-loop, 3
Unitarity, 4, 107

Vacuum, 10
– physical, 2
Vacuum expectation values, 15
Vacuum polarization, 2, 3, 17, 37, 48
Vacuum-vacuum amplitude, 10
Van Fleck–Morette determinant, 15, 25, 29, 32
Variation of gauge, 94, 95
Vertex functions, 11, 12
– exact, 1
– of perturbation theory, 2
Volume element, 20
– of configuration space, 13

Ward identities, 94, 99
Wave equation, 3, 51
Weyl curvature tensor, 85
World function, 21, 24, 25

Zero eigenvalue, 24, 34
Zero modes, 116, 119, 120, 122
– conformal, 111
Zero-charge, 5, 102, 103, 105, 107
Zeta-function, 7, 116–119

Printing: Weihert-Druck GmbH, Darmstadt
Binding: Buchbinderei Schäffer, Grünstadt

Lecture Notes in Physics

For information about Vols. 1–508
please contact your bookseller or Springer-Verlag

Vol. 509: J. Wess, V. P. Akulov (Eds.), Supersymmetry and Quantum Field Theory. Proceedings, 1997. XV, 405 pages. 1998.

Vol. 510: J. Navarro, A. Polls (Eds.), Microscopic Quantum Many-Body Theories and Their Applications. Proceedings, 1997. XIII, 379 pages. 1998.

Vol. 511: S. Benkadda, G. M. Zaslavsky (Eds.), Chaos, Kinetics and Nonlinear Dynamics in Fluids and Plasmas. Proceedings, 1997. VIII, 438 pages. 1998.

Vol. 512: H. Gausterer, C. Lang (Eds.), Computing Particle Properties. Proceedings, 1997. VII, 335 pages. 1998.

Vol. 513: A. Bernstein, D. Drechsel, T. Walcher (Eds.), Chiral Dynamics: Theory and Experiment. Proceedings, 1997. IX, 394 pages. 1998.

Vol. 514: F. W. Hehl, C. Kiefer, R. J. K. Metzler, Black Holes: Theory and Observation. Proceedings, 1997. XV, 519 pages. 1998.

Vol. 515: C.-H. Bruneau (Ed.), Sixteenth International Conference on Numerical Methods in Fluid Dynamics. Proceedings. XV, 568 pages. 1998.

Vol. 516: J. Cleymans, H. B. Geyer, F. G. Scholtz (Eds.), Hadrons in Dense Matter and Hadrosynthesis. Proceedings, 1998. XII, 253 pages. 1999.

Vol. 517: Ph. Blanchard, A. Jadczyk (Eds.), Quantum Future. Proceedings, 1997. X, 244 pages. 1999.

Vol. 518: P. G. L. Leach, S. E. Bouquet, J.-L. Rouet, E. Fijalkow (Eds.), Dynamical Systems, Plasmas and Gravitation. Proceedings, 1997. XII, 397 pages. 1999.

Vol. 519: R. Kutner, A. Pękalski, K. Sznajd-Weron (Eds.), Anomalous Diffusion. From Basics to Applications. Proceedings, 1998. XVIII, 378 pages. 1999.

Vol. 520: J. A. van Paradijs, J. A. M. Bleeker (Eds.), X-Ray Spectroscopy in Astrophysics. EADN School X. Proceedings, 1997. XV, 530 pages. 1999.

Vol. 521: L. Mathelitsch, W. Plessas (Eds.), Broken Symmetries. Proceedings, 1998. VII, 299 pages. 1999.

Vol. 522: J. W. Clark, T. Lindenau, M. L. Ristig (Eds.), Scientific Applications of Neural Nets. Proceedings, 1998. XIII, 288 pages. 1999.

Vol. 523: B. Wolf, O. Stahl, A. W. Fullerton (Eds.), Variable and Non-spherical Stellar Winds in Luminous Hot Stars. Proceedings, 1998. XX, 424 pages. 1999.

Vol. 524: J. Wess, E. A. Ivanov (Eds.), Supersymmetries and Quantum Symmetries. Proceedings, 1997. XX, 442 pages. 1999.

Vol. 525: A. Ceresole, C. Kounnas, D. Lüst, S. Theisen (Eds.), Quantum Aspects of Gauge Theories, Supersymmetry and Unification. Proceedings, 1998. X, 511 pages. 1999.

Vol. 526: H.-P. Breuer, F. Petruccione (Eds.), Open Systems and Measurement in Relativistic Quantum Theory. Proceedings, 1998. VIII, 240 pages. 1999.

Vol. 527: D. Reguera, J. M. G. Vilar, J. M. Rubí (Eds.), Statistical Mechanics of Biocomplexity. Proceedings, 1998. XI, 318 pages. 1999.

Vol. 528: I. Peschel, X. Wang, M. Kaulke, K. Hallberg (Eds.), Density-Matrix Renormalization. Proceedings, 1998. XVI, 355 pages. 1999.

Vol. 529: S. Biringen, H. Örs, A. Tezel, J.H. Ferziger (Eds.), Industrial and Environmental Applications of Direct and Large-Eddy Simulation. Proceedings, 1998. XVI, 301 pages. 1999.

Vol. 530: H.-J. Röser, K. Meisenheimer (Eds.), The Radio Galaxy Messier 87. Proceedings, 1997. XIII, 342 pages. 1999.

Vol. 531: H. Benisty, J.-M. Gérard, R. Houdré, J. Rarity, C. Weisbuch (Eds.), Confined Photon Systems. Proceedings, 1998. X, 496 pages. 1999.

Vol. 532: S. C. Müller, J. Parisi, W. Zimmermann (Eds.), Transport and Structure. Their Competitive Roles in Biophysics and Chemistry. XII, 400 pages. 1999.

Vol. 533: K. Hutter, Y. Wang, H. Beer (Eds.), Advances in Cold-Region Thermal Engineering and Sciences. Proceedings, 1999. XIV, 608 pages. 1999.

Vol. 534: F. Moreno, F. González (Eds.), Light Scattering from Microstructures. Proceedings, 1998. XII, 300 pages. 2000

Vol. 535: H. Dreyssé (Ed.), Electronic Structure and Physical Properties of Solids: The Uses of the LMTO Method. Proceedings, 1998. XIV, 458 pages. 2000.

Vol. 536: T. Passot, P.-L. Sulem (Eds.), Nonlinear MHD Waves and Turbulence. Proceedings, 1998. X, 385 pages. 1999.

Vol. 537: S. Cotsakis, G. W. Gibbons (Eds.), Mathematical and Quantum Aspects of Relativity and Cosmology. Proceedings, 1998. XII, 251 pages. 1999.

Vol. 538: Ph. Blanchard, D. Giulini, E. Joos, C. Kiefer, I.-O. Stamatescu (Eds.), Decoherence: Theoretical, Experimental, and Conceptual Problems. Proceedings, 1998. XII, 345 pages. 2000.

Vol. 539: A. Borowiec, W. Cegła, B. Jancewicz, W. Karwowski (Eds.), Theoretical Physics. Fin de Siècle. Proceedings, 1998. XX, 319 pages. 2000.

Vol. 540: B. G. Schmidt (Ed.), Einstein's Field Equations and Their Physical Implications. Selected Essays. 1999. XIII, 429 pages. 2000

Vol. 541: J. Kowalski-Glikman (Ed.), Towards Quantum Gravity. Proceedings, 1999. XII, 376 pages. 2000.

Vol. 542: P. L. Christiansen, M. P. Sørensen, A. C. Scott (Eds.), Nonlinear Science at the Dawn of the 21st Century. Proceedings, 1998. XXVI, 458 pages. 2000.

Vol. 543: H. Gausterer, H. Grosse, L. Pittner (Eds.), Geometry and Quantum Physics. Proceedings, 1999. VIII, 408 pages. 2000.

Vol. 545: J. Klamut, B. W. Veal, B. M. Dabrowski, P. W. Klamut, M. Kazimierski (Eds.), New Developments in High-Temperature Superconductivity. Proceedings, 1998. VIII, 275 pages. 2000.

Vol. 546: G. Grindhammer, B. A. Kniehl, G. Kramer (Eds.), New Trends in HERA Physics 1999. Proceedings, 1999. XIV, 460 pages. 2000.

Monographs
For information about Vols. 1–20
please contact your bookseller or Springer-Verlag

Vol. m 21: G. P. Berman, E. N. Bulgakov, D. D. Holm, Crossover-Time in Quantum Boson and Spin Systems. XI, 268 pages. 1994.

Vol. m 22: M.-O. Hongler, Chaotic and Stochastic Behaviour in Automatic Production Lines. V, 85 pages. 1994.

Vol. m 23: V. S. Viswanath, G. Müller, The Recursion Method. X, 259 pages. 1994.

Vol. m 24: A. Ern, V. Giovangigli, Multicomponent Transport Algorithms. XIV, 427 pages. 1994.

Vol. m 25: A. V. Bogdanov, G. V. Dubrovskiy, M. P. Krutikov, D. V. Kulginov, V. M. Strelchenya, Interaction of Gases with Surfaces. XIV, 132 pages. 1995.

Vol. m 26: M. Dineykhan, G. V. Efimov, G. Ganbold, S. N. Nedelko, Oscillator Representation in Quantum Physics. IX, 279 pages. 1995.

Vol. m 27: J. T. Ottesen, Infinite Dimensional Groups and Algebras in Quantum Physics. IX, 218 pages. 1995.

Vol. m 28: O. Piguet, S. P. Sorella, Algebraic Renormalization. IX, 134 pages. 1995.

Vol. m 29: C. Bendjaballah, Introduction to Photon Communication. VII, 193 pages. 1995.

Vol. m 30: A. J. Greer, W. J. Kossler, Low Magnetic Fields in Anisotropic Superconductors. VII, 161 pages. 1995.

Vol. m 31 (Corr. Second Printing): P. Busch, M. Grabowski, P.J. Lahti, Operational Quantum Physics. XII, 230 pages. 1997.

Vol. m 32: L. de Broglie, Diverses questions de mécanique et de thermodynamique classiques et relativistes. XII, 198 pages. 1995.

Vol. m 33: R. Alkofer, H. Reinhardt, Chiral Quark Dynamics. VIII, 115 pages. 1995.

Vol. m 34: R. Jost, Das Märchen vom Elfenbeinernen Turm. VIII, 286 pages. 1995.

Vol. m 35: E. Elizalde, Ten Physical Applications of Spectral Zeta Functions. XIV, 224 pages. 1995.

Vol. m 36: G. Dunne, Self-Dual Chern-Simons Theories. X, 217 pages. 1995.

Vol. m 37: S. Childress, A.D. Gilbert, Stretch, Twist, Fold: The Fast Dynamo. XI, 406 pages. 1995.

Vol. m 38: J. González, M. A. Martín-Delgado, G. Sierra, A. H. Vozmediano, Quantum Electron Liquids and High-Tc Superconductivity. X, 299 pages. 1995.

Vol. m 39: L. Pittner, Algebraic Foundations of Non-Com-mutative Differential Geometry and Quantum Groups. XII, 469 pages. 1996.

Vol. m 40: H.-J. Borchers, Translation Group and Particle Representations in Quantum Field Theory. VII, 131 pages. 1996.

Vol. m 41: B. K. Chakrabarti, A. Dutta, P. Sen, Quantum Ising Phases and Transitions in Transverse Ising Models. X, 204 pages. 1996.

Vol. m 42: P. Bouwknegt, J. McCarthy, K. Pilch, The W3 Algebra. Modules, Semi-infinite Cohomology and BV Algebras. XI, 204 pages. 1996.

Vol. m 43: M. Schottenloher, A Mathematical Introduction to Conformal Field Theory. VIII, 142 pages. 1997.

Vol. m 44: A. Bach, Indistinguishable Classical Particles. VIII, 157 pages. 1997.

Vol. m 45: M. Ferrari, V. T. Granik, A. Imam, J. C. Nadeau (Eds.), Advances in Doublet Mechanics. XVI, 214 pages. 1997.

Vol. m 46: M. Camenzind, Les noyaux actifs de galaxies. XVIII, 218 pages. 1997.

Vol. m 47: L. M. Zubov, Nonlinear Theory of Dislocations and Disclinations in Elastic Body. VI, 205 pages. 1997.

Vol. m 48: P. Kopietz, Bosonization of Interacting Fermions in Arbitrary Dimensions. XII, 259 pages. 1997.

Vol. m 49: M. Zak, J. B. Zbilut, R. E. Meyers, From Instability to Intelligence. Complexity and Predictability in Nonlinear Dynamics. XIV, 552 pages. 1997.

Vol. m 50: J. Ambjørn, M. Carfora, A. Marzuoli, The Geometry of Dynamical Triangulations. VI, 197 pages. 1997.

Vol. m 51: G. Landi, An Introduction to Noncommutative Spaces and Their Geometries. XI, 200 pages. 1997.

Vol. m 52: M. Hénon, Generating Families in the Restricted Three-Body Problem. XI, 278 pages. 1997.

Vol. m 53: M. Gad-el-Hak, A. Pollard, J.-P. Bonnet (Eds.), Flow Control. Fundamentals and Practices. XII, 527 pages. 1998.

Vol. m 54: Y. Suzuki, K. Varga, Stochastic Variational Approach to Quantum-Mechanical Few-Body Problems. XIV, 324 pages. 1998.

Vol. m 55: F. Busse, S. C. Müller, Evolution of Spontaneous Structures in Dissipative Continuous Systems. X, 559 pages. 1998.

Vol. m 56: R. Haussmann, Self-consistent Quantum Field Theory and Bosonization for Strongly Correlated Electron Systems. VIII, 173 pages. 1999.

Vol. m 57: G. Cicogna, G. Gaeta, Symmetry and Perturbation Theory in Nonlinear Dynamics. XI, 208 pages. 1999.

Vol. m 58: J. Daillant, A. Gibaud (Eds.), X-Ray and Neutron Reflectivity: Principles and Applications. XVIII, 331 pages. 1999.

Vol. m 59: M. Kriele, Spacetime. Foundations of General Relativity and Differential Geometry. XV, 432 pages. 1999.

Vol. m 60: J. T. Londergan, J. P. Carini, D. P. Murdock, Binding and Scattering in Two-Dimensional Systems. Applications to Quantum Wires, Waveguides and Photonic Crystals. X, 222 pages. 1999.

Vol. m 61: V. Perlick, Ray Optics, Fermat's Principle, and Applications to General Relativity. X, 220 pages. 2000.

Vol. m 63: R. J. Szabo, Ray Optics, Equivariant Cohomology and Localization of Path Integrals. XII, 315 pages. 2000.

Vol. m 64: I. G. Avramidi, Heat Kernel and Quantum Gravity. X, 143 pages. 2000.